Annual International Conference on Data Science, Machine Learning and Blockchain Technology (AICDMB 2021)

IOP Conference Series: Materials Science and Engineering
Volume 1110

Mysuru, India
15 - 16 February 2021

Editors:

H.L. Gururaj
V. Ravikumar
M.C. Lavanya

ISBN: 978-1-7138-3400-7
ISSN: 1757-8981

Printed from e-media with permission by:

Curran Associates, Inc.
57 Morehouse Lane
Red Hook, NY 12571

Some format issues inherent in the e-media version may also appear in this print version.

Institute of Physics
Dirac House, Temple Back
Bristol BS1 6BE UK

Phone: 44 1 17 929 7481
Fax: 44 1 17 920 0979

techtracking@iop.org

Additional copies of this publication are available from:

Curran Associates, Inc.
57 Morehouse Lane
Red Hook, NY 12571 USA
Phone: 845-758-0400
Fax: 845-758-2633
Email: curran@proceedings.com
Web: www.proceedings.com

TABLE OF CONTENTS

IOP Conf. Series: Materials Science and Engineering 1110 (2021) 011001 doi:10.1088/1757-899X/1110/1/011001

Preface

It is our pleasure to welcome you to the 2021 2nd Annual International Conference on Data Science Machine Learning and Blockchain Technology (AICDMB 2021) in Mysuru, Karnataka. This conference aims to bring researchers, scientists, engineers, and industrialists from all over the world to discuss and share advances and challenges in digital transformation.

We hope you will have a technically rewarding experience, and this wide platform provides an opportunity to enhance your skills.

AICDMB 2021 promises to be both stimulating and informative with a wonderful array of keynote and invited speakers from all over the world. Delegates will have a wide range of sessions to choose from and will have a difficult decision in deciding which sessions to attend.

The proceedings are a compilation of all the accepted papers and represent an interesting outcome of the conference. We would like to acknowledge all of those who supported AICDMB 2021. Each individual and institutional help were very important for the success of this conference. Specially we would like to extend our gratitude to the organizing committee for their valuable advice in the organization of this conference.

We sincerely hope that AICDMB2021 will be a forum for excellent discussions that will put forward new ideas and promote collaborative research. The proceedings will serve as an important research source of references and the knowledge.

IOP Conf. Series: Materials Science and Engineering 1110 (2021) 011001 doi:10.1088/1757-899X/1110/1/011001

Editors

Dr. Gururaj H L
ACM Distinguished Speaker & Associate Professor
Program co-chair (AICDMB-2021)
Computer Science & Engineering
Vidyavardhaka College of Engineering, Mysuru, India

Dr. Ravikumar V
Professor & Head
Program chair (AICDMB-2021)
Computer Science & Engineering
Vidyavardhaka College of Engineering, Mysuru, India

Prof. Lavanya M C
Assistant Professor
Computer Science & Engineering
Vidyavardhaka College of Engineering, Mysuru, India

IOP Conf. Series: Materials Science and Engineering 1110 (2021) 011001 doi:10.1088/1757-899X/1110/1/011001

Organizing Committee of Conference

General Chairs & Co-chairs

Dr. B Sadashive Gowda
Dr. G B Krishnappa

Program Chairs & Co-chairs

Dr. Ravi Kumar V
Dr. Gururaj H L

Publication Co-chairs

Dr. Ramakrishna Hegde
Prof. Pooja M R
Prof. Anusha K S

Publicity Co-chairs

Dr. Aditya C R
Prof. Pavan Kumar S P
Prof. Nithin Kumar

Web Co-chairs

Prof. Natesh M
Prof. Rakesh K R
Prof. Shashank N

Sponsorship Co-chairs

Prof. Janhavi V
Prof. Swathi B H

Registration Co-chairs

Prof. Hamsaveni M
Prof. Harshitha K
Prof. Chayakumari H A
Prof. Pruthvi P R

Finance Co-chairs

Dr. Paramesha K
Prof. Ravi P
Prof. M C Lavanya

IOP Conf. Series: Materials Science and Engineering 1110 (2021) 011002 doi:10.1088/1757-899X/1110/1/011002

Peer review declaration

All papers published in this volume of *IOP Conference Series: Materials Science and Engineering* have been peer reviewed through processes administered by the Editors. Reviews were conducted by expert referees to the professional and scientific standards expected of a proceedings journal published by IOP Publishing.

- **Double-blind**

- **Describe criteria used by Reviewers when accepting/declining papers. Was there the opportunity to resubmit articles after revisions?**

  The criteria used for review process are:
  Organization of the paper
  Originality/Novelty
  Relevance to the Field(s) of Conference
  Technical quality of the paper
  Clarity of Presentation

  Yes, based on the review comments the paper should be revised before submitting the final manuscript.

- **Conference submission management system: Easy-chair**

- **Number of submissions received: 53**

- **Number of submissions sent for review: 50**

- **Number of submissions accepted:25**

- **Acceptance Rate (Number of Submissions Accepted / Number of Submissions Received X 100):47.16**

- **Average number of reviews per paper: 02**

- **Total number of reviewers involved:35**

- **Any additional info on review process (ie plagiarism check system):**

  Turnitin Plagiarism Tool

- **Contact person for queries: Dr. Gururaj H L (gururaj1711@vvce.ac.in)**

Blockchain for Healthcare Sector- Analytical Review

Nail Adeeb Ali Abdu[1] and Zhaoshun Wang[2]

[1,2] Department of Computer Science and Technology, School of Computer and Communication Engineering, University of Science and Technology Beijing (USTB), Beijing 10083, China; zhswang@ustb.edu.cn

[1]**E-mail:** b20190665@xs.ustb.edu.cn , n3a_alkershi@yahoo.com

Abstract. Blockchain is a collection of records and blocks that are integrated using cryptography. Every block consists of certain levels of cryptographic hash for the previous block in the form of a timestamp and the transaction data integral to the transaction record. The blockchain systems have currently used among the various set of industrial information networks. If effectively used, it shall help the system in having more secured levels of development. In this research review, the objective is to understand the gamut of blockchain implementation and to pursue the scope in the usage of the blockchain solutions in healthcare solutions. Review of the literature and data from the case scenario analysis throws light on a certain set of constraints and vulnerabilities of using the blockchain in healthcare which needs attention. Profoundly the constraints are about the quality of nodes verifying the blocks and the anomaly access of information from the blocks. By overcoming such constraints, there will be an improvement of effectiveness with which the system gets implemented.

Keywords: Cyber-attack, data Management, Blockchain, Healthcare, Personalized medicine, Gap Analysis.

1.Introduction

Information and communication technologies are evolving so rapidly that there is a paradigm shift in the way businesses function, the way transactions and sensitive information managed across the verticals. Among the contemporary range of solutions that have evolved in the ICT, blockchain is one of the new age developments, which increased the veracity and security of data management. The trend of increasing cyber-attacks on data and the prevailing information security concerns necessitate for a set of data security solutions in place, which makes a significant impact on the companies [1].

Information systems security has significant importance as any information breach might be a very challenging issue for many stakeholders associated with the system. For instance, in the case of the banking system or digital transactions, there are a distinct set of challenges[2]. Among the first set challenges, ensuring the data security and data integrity are predominantly important. In the case of the data security, it is about ensuring there is an appropriate kind of firewall systems, anomaly detection systems etc. Whereas in the case of data integrity, it is profoundly about securing the data from any manipulation to the records and mitigating the risks of data security[3].

Blockchain, overall, can be defined as a system that ensures the transaction data is secured and monitored for amendments or trans positioning of data from one level to the other. Such system shall help the system in having more secured levels of development [1]. In this research proposal, the objective is to understand the gamut of blockchain implementation and to pursue the scope in the usage of the blockchain solutions in healthcare solutions.

In the further sections of this report, the emphasis is on the conceptual outline of the blockchain solutions and how the blockchain solutions used across industrial verticals. Further, it follows by a review of literature on the blockchain implementation in the healthcare industry. The further section discusses the gap in the existing system implementation and the potential scope for future research towards improvising contemporary solutions.

2.Blockchain System

Blockchain is a collection of records and blocks that are integrated using the cryptography. Every block consists of certain levels of cryptographic hash for the previous block in the form of a timestamp and the transaction data integral to the transaction record[3]. In terms of design, the blockchain can be more significant in terms of curtailing the chances for data or transaction input modification. It functions as a distributed ledger which can work on transaction records among the parties, making it efficient and effective, in terms of verification and holding the records in structured formats [4].

It is imperative from the literature that the blockchain is undeniably high levels of ingenious development. Today, across the verticals, the role of blockchain solutions is becoming an impeccable development. The crux of blockchain solutions is about the scope for disclosure of the transaction input public to the domain, but do not allow any changes to the transactions[5].

The blockchain model was introduced for cryptocurrency solutions initially. Over time the model of blockchain garnered the attention of other financial verticals, and today the blockchain solutions are more prevalent across a distinct set of industrial verticals [6].

Information managed on the blockchains exists as a shared and reconciled set of database. It supports the network with numerous benefits, as the blockchains are usually not stored in one single location and the data managed as truly public and verifiable by the stakeholders. As there is a lack of centralized versions for the information, it shall be very challenging for the hackers to corrupt the data. The other advantage is about how the blockchain is hosted across multiple computers in a similar instance, which makes it easier in terms of ease of access [7].

2.1Instance of Blockchain

In a illustrative scenario, it can be stated that when there is need for documents to be reviewed and edited by the teams in an organization with in quick turnaround time, the traditional way of sharing the documents in collaborative networks like cloud or share point and the review in terms of modifications to the model has to undergo only at one source. Such structure is more prevalent in the databases of the banking and financial systems and other such verticals [5].

The way many cloud-based document or transaction systems exists in the current trend is supporting the organizations in having more effective improvisations, but the challenge exists when millions of records must be managed in terms of transaction records.

In the case of the transaction records, when there is too much of back and forth of data, it leads to time lag and redundancy in terms of records. But with the blockchain the authenticated review of the records can take place in real-time for the systems.

Some of the key reasons why the blockchain has gained significance in the domain are:

- The process is not owned by any single entity and it is profoundly decentralized.
- Cryptographic ways of managing the records are among the critical advantage in the system.
- Block chain is immutable and the scope in terms of tampering the data in the blockchain system is not feasible.
- Transparency in the system is very high and it shall help the organizations improve the overall solution outlook [4]

Management of the blockchain systems is predominantly handled by peer-to-peer network in the form of collection of nodes that are interconnected to the other. The nodes could be individual systems or the respective information systems that support in information exchange [7]. The blockchain system embraces the network of peer-to-peer network wherein the partitions in the entire workload is distributed among the participants, recognized as peers to the system. There is hardly any kind of central servers and the structure is formation of distributed and decentralized peers. Such system helps in improving the overall ways of information management and towards developing a comprehensive information network [6].

3.Blockchain Usage across Sectors

Blockchain solution mitigates the intervention of the intermediary inputs for the transactions. Personal computing has become an integral to the public usage by the respective stakeholders, by the interface created using the GUI, which is technically defined as wallet applications, which people shall use for the transactions. Transactions are usually connected by the processes of identity verification. It is pragmatic that the wallet apps shall transform into identity management solutions [8].

Though in the earlier days, the usage of the blockchain solutions was more prevalent in the cryptocurrency market, today, there is ongoing trends of embracing the blockchain kind of solutions to various industrial verticals. Categorically, in the banking, financial and insurance domain, the implementation of blockchain based transaction processing systems are yielding effective outcome, thus encouraging more other verticals to embrace the solution [9].

Some of the key areas wherein the blockchain solutions are vividly used are:

3.1 BFSI

Banking, finance and insurance industry is among the front runners in the adoption of block chain system, post its success in the cryptocurrency market. Considering the volume of transactions and the intrinsic need for security and efficiency in information sharing among the groups, the usage of blockchain solutions in the BFSI domain has become significant and there are many companies that are experiencing the results of such integral usage [6].

3.2 Governance Systems
In many of the developed and developing countries, there are departments across the services, using the solutions like the blockchain. For instance, in the Indian government, there are many departments like the public works, judiciary department, land and infrastructure departments and many more are focusing on the usage of the blockchain solutions. Such holistic and integrated ways of developing an eco-system of information sharing shall support the organizations and stakeholders with more sustainable outcome [7].

3.3 Stock Trading
One of the critical areas wherein the role of blockchain solutions is considered significant in the stock trading domain. The potential in terms of adding the efficiency in share settlements turns to be strong use case for the blockchains in the trading industry [5]. The process of peer to peer trade confirmation models helps the parties in instantaneous settlements of the trade. Many intermediary processes like the clearing houses, custodians and the auditor's role in the transaction processing is mitigated, thus enabling more quicker solutions in the transaction processing [10].

3.4 Supply Chain Auditing
Blockchain solutions usage in the case of the supply chain audits are turning to be potential developments which is improving the scope of reducing the pilferage or any scope for fault in delivery systems. With improved systems of instant sharing of information among the supply chain partners network, it becomes much easier for the organizations towards ensuring right kind of developments in the process [8].

3.5 Identity and KYC management
The other critical area wherein the blockchains solution is prevalently used is the identity and KYC management. For instance, the Know Your Customer records are very important and is essential across various service provider networks. When the system is integrated on the blockchain platforms, it becomes much easier for the system shareholders to have adequate and appropriate kind of information set, which will mitigate the risk of tampering the data and towards holistic improvement in the system [11].

3.6 Healthcare Systems
Blockchain systems usage in healthcare is one of the key developments that is turning to be more resourceful for the organizations. For instance, in the implementation of block chain model for managing the sensitive data of the patients and towards having easy access to updated and real-time information, it helps in quick access to patient data [2]. The medical professionals can have more significant information of patient's health conditions and are able to plan the medical services in comprehensive manner.

4. Blockchain in Healthcare

Blockchain is gaining popularity in the healthcare industry as the age-old issues of the industry can be effectively addressed in the system implementation. Ineffective handling of the data, higher cost of data coordination among the stakeholders of the industry etc. are turning to be a challenging condition for the organizations. There are many healthcare companies eying on the blockchain solutions as a potential option towards addressing the limitations in the existing processes [3].

Key barrier in the healthcare system is about the data sharing and the transformation towards value-based care, wherein the critical role for the data sharing among stakeholders are very vital [12]. There is considerable level of silos in the current system for the companies and there is weaker section of information exchange among the groups and if the system is rightly integrated, it will help in connecting the silos and enabling optimum sharing and utilization of the resources, which can help in comprehensive development [6].

Interoperability is one of the critical issues for the healthcare industry. If the issue is addressed with effective integration and coordination among the stakeholders, it shall ease the process complexities and lead to more integrated approach in the system implementation [5].

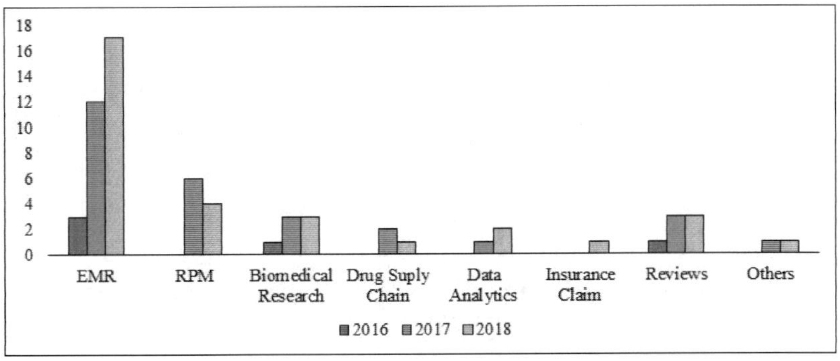

Figure 4.1. Blockchain usage in various divisions of healthcare management [9].

There are vivid areas wherein the blockchain solutions in the healthcare can be potential solution. Few of the case dynamics that are imperative as per the literature are:

4.1 Electronic Health Records

One of the key areas wherein the solutions can be more robust for the organizations are the electronic health records solutions. In the case of the systems being in silos, the information of the patients is with different entities and in distinct formats. In the case of trauma care or emergency medical services, the scope of collating the distributed information towards historic information analysis of the patient becomes a challenge [8].

In a illustrative scenario, if the medical records of the patient are in different medical institution, and the current trauma care center needs access to such data or historic information of the earlier treatments to a patient, it might not be available, which turns to be a potential challenge [6]. But when there is integrated system, wherein the information is exchanged over the distributed

IOP Conf. Series: Materials Science and Engineering 1110 (2021) 012001 doi:10.1088/1757-899X/1110/1/012001

ledger model, there is real-time access to the information, which can lead to sustainable kind of development process [13].

4.2 Third party information management

In addition to the medical professionals and the healthcare institutions, there are many third parties too who are integral to the healthcare industry. For instance, when the patient gets hurt in an accident and is admitted to trauma care center, there are certain records that are essential for the police department (or in certain instances the police might update information in their records about the cause of injury or other such kind of information) which might have significant impact on the conditions of treatment [6].

The other dimension is about the insurance companies processing the requests for insurance approvals and claims. If there is blockchain integrated data sharing system, there are potential ways in which the data could be more effectively managed among the systems [9].

In the presence of holistic system towards managing the blockchain solutions, there is imperative scope for instantaneous sharing of information among the stakeholders and the ones that could support in improving the kind of overall sustainable development model.

There are many case scenarios that are reviewed in the literature, pertaining to block chain-based healthcare system. Few organizations are currently engaged in and structured format of managing the block chain-based system for improving the healthcare for professionals and patients too. If the patient health history information is decentralized, it will lead to more comprehensive levels of service management. Right from health history tracking to claims management, fraudulent claims processing, classified information sharing for medical research, there are many areas in which the solutions can be more pragmatic [4].

A medical system developed on the Ethereum block chain as MedRec is a contemporary solution that prioritizes the patient record information to the agencies and enables transparent and accessible levels of view of medical data, thus making it simpler for the patients and doctors for purview. To ensure secured access to the data, the system works based on the Proof of Authority Mechanism.

4.3 Potential areas wherein Block Chain Solutions can be used

The blockchain solutions in the healthcare is currently used in the case of the medical records management, claims processing in the insurance for the healthcare, transaction records processing etc. Also, in many nations wherein the government schemes are prevalent for healthcare, towards managing the record of services to the patients, the blockchain based solution system can be more pragmatic.

As discussed above, there are phenomenal ways in which the blockchain based healthcare solutions can be more emphatic for potential healthcare services.

4.4 Integrated System of Patient Records

Longitudinal records of patients, with the structure of managing information like the episodes of treatment, disease registries, diagnostic report information, path lab reports, and the treatment schedules can be combined from different sources in to one system, in the blockchain environment. This kind of holistic information sharing and management among the key stakeholders shall support in quick response of treatment to the patient [9].

4.5 Master Patient Indices

Having a collaborated and one structure of data management shall ease the complexities of mismatch or redundancy in the data management, which shall support in improving the overall systems and practices that are integral to development.

4.6 Supply Chain Management

Among the healthcare companies, there is integral system for robust supply chain process. If the patient data and the inventory data in the healthcare institutions are managed effectively with a block chain solution, it will help in mitigating the pilferage and ensuring there is more systematic facets of handling the supply chain solutions [5].

4.7 Improving the Transparency and Security

Two of the critical factors in the healthcare data is about ensuring the security of the patient data and the other factor is about transparency in the system. In the case of the blockchain based models, the transparency and security can be realized if the right kind of system is embraced for implementation [8].

The Figure 4.2 below indicates the scope of how the blockchain systems are used in the healthcare domain and the key areas wherein the solution is highly used for operational processes.

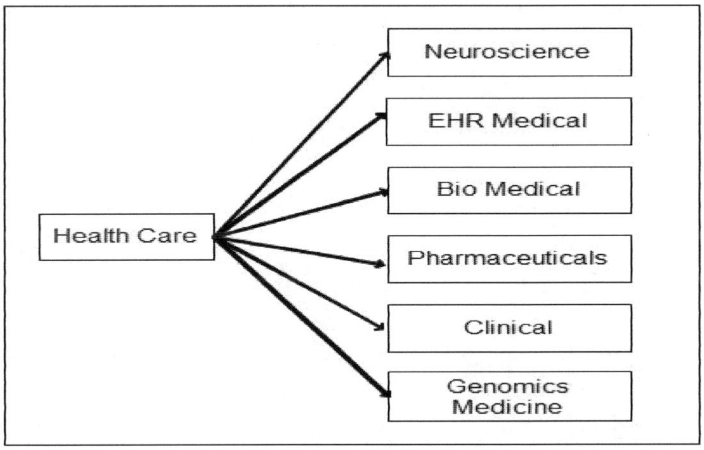

Figure 4.2. Application of blockchain technology in healthcare [14].

5. Review of Literature

The work researched in the domain of blockchain implementation in healthcare. Profoundly the study is a review of the literature discussed in the earlier studies. The authors of the study indicate that there are distinct ways in which the research studies highlight the benefits and the cons of the system, but very few studies have focused on the live data conditions pertaining to how the block chain in healthcare can work. But the authors reiterate the scope and efficacy of the system towards more emphatic levels of medical records management[15].

In [14], the authors the study too emphasize the need for more structured research in the domain, as predominantly many of the case studies discussed in the process indicates about the usage of the system. There are distinct set of conditions that requires the verification of the wallet application systems. Though there are prototypes and case scenario presentations, still in terms of creating significant system, there is need for better process and proven conditions for testing the application in healthcare domain[14].

In [16], the study has focused on the dimension of customization required for the healthcare systems. Depending on the conditions and the contextual implementation of the systems, the scope for sharing of the data in the healthcare systems might vary. It is paramount importance that such customization requirements are to be considered with the implementation of blockchain system[16].

The study highlights the fact that the blockchain system though creates complexities for hacking, still in terms of security, there are certain vulnerabilities and concerns that must be addressed in order to make it a potential and fool-proof system.

In [17], the authors of the study too have taken up the conditions of implementing the blockchain based system and has highlighted about five critical factors that make significant difference to the system implementation[17]. Firstly, the scope of digital access rules, followed by data aggregation, liquidity, identity management and the scope of data immutability. If the systems are effectively screened and customized for the aforementioned factors, there are potential chances that the system can be more robust and significant to the developments.

In [18], the focus of the study is about using the IT based framework structures that can support in integrated system implementation. The study discusses profoundly about the scope of using the IoT systems as a catalyst towards managing the transaction process and improve the process management quality of healthcare systems[18].

The authors of the study [19] has focused on the "Shared Nationwide Interoperability Roadmap" from the Office of the National Coordinator for Health Information Technology (ONC) and has evaluated the conditions that drive to the requirements of how the system has to be more robust and the models that are to be considered towards significant improvement to the systems. The study has discussed the crux factor of blockchain system, the decentralization as the structural framework that could be implemented as per the requirements that are to be addressed for the ONC system of managing the healthcare data[19].

Pseudonym Based Encryption with Different Authorities (PBE-DA) is proposed in [13], wherein PBE-DA will be resourceful for the patient anonymously to access, check or update their sensitive data on EHRs system. The system discusses more about the IoT kind of system integration that can support in improving the quality and access to the data for the organizations[13].

In [20], the study has focused on how effective the blockchain system could be in enabling the transition of the datacenter from institution centric to the patient data centric conditions. Highlighting certain set of limitations in the blockchain system implementation, the study triggers the points towards understanding the key factors that can be integral to ensuring adequate and appropriate kind of system development[10].

In [20], the study highlights the scope of blockchain solution implementation towards understanding the developments in the domain and how effectively the model can be resourceful for implementing more integrated and transparent system in the biomedical and healthcare solutions[20], [9].

5.1 Blockchains in Personalized Medicine

Personalized medicine can be broadly defined as the classification of patients as a group based on indicators like genomic data, race, age, or gender and is an evolution of EHR/EMRs. Treatment, health plans and relevant advertising are developed based on classification as a group. Personalized medicine data faces a few issues from two aspects (i) ethical (ii) and access point of view [21], [22]. The ethical point of view raises concerns on individual data privacy and the family genome exposed which might not be desired by the entire family members. Further on access if the data is used for another study, it needs the patient/family consent on privacy aspect the approval of which might take the time or may turn away the patient/family from availing the healthcare. It might also lead to a case where the fear of data exposure might lead to false information supplied by the patient/family.

The integration of blockchain into the personalized medicine model may address the issues as mentioned earlier in data aspects. The feature of encryption will ensure that the private information is accessed by select few giving the patient/family confidence. Also, the record of access and notifications received will keep the patient/family informed. On the data end-user front, the doctors or researchers can access the data in a faster way reducing waiting time for approvals. This two-way seamless integration enables the patient/family to donate or sell their data for experiments or to other patients. The actual data stored in data lakes can be from clinical trials or clinical records but goes through authentication and decryption before final disclosure. The disclosure is limited to links to the data. These models can be perused in MedRec [23] and the Stony Brook Oncology project [24] and [25]. The pharmaceutical and insurance companies also can be end-users in these models.

5.2 Major Implementations

The idea to use blockchains in healthcare systems started in 2016. The use of blockchains for cryptocurrencies and the model of distributed ledger and better access control paved the way to replicate the model in healthcare systems. Few implementations described below:

Massachusetts Institute of Technology (MIT) developed a blockchain framework of healthcare system named MedRec [23]. The model focused on four aspects: data fragmentation, interoperability, patient focus, and medical diagnosis data.

Further it proposed contracts for managing data queries and linking between the patient and the caretaker. The drawbacks identified in this system are end-point security, flexibility, and patient intervention in the blockchain. The MedRec 2.0 [26] model in 2018 addressed patient inference from transactions and some scalability and privacy issues. But the end-point security remains a challenge.

In 2018 a model Medical Chain [27] was introduced with patient centricity in the management of EHRs. The key feature was the use of Hyper Ledger Fabric for access control. Along with access control to the patient the model provided a marketplace for negotiation with third parties for data access. MedTokens are introduced for patients to be purchased and stored. The drawbacks noticed were non-purchase of tokens and lack of legal qualification for MedToken security. Patientory [28] is a similar model like Medical Chain [27].

Along with tokens, the key aspect is privacy and security of patient are secured by HIPAA-compliant data storage systems. The tokens on this platform can be used for storage space or third party contract functions. In [29], [30] it is observed that privacy is an issue using blockchains Internet-of-Things (IoT) space. The user identities are observed to be inferred in the network despite encryption by monitoring device activity giving scope for attack by malicious users.

In [31], [30] the existing cloud-based healthcare systems are integrated with blockchains. [31] focused on a hypothesis of whether an existing cloud based EHR/EMR system could be built upon with improved data security with blockchains. The drawback observed in this model is complying to the privacy laws and regulations. In [30], the key focus is pseudonymity (username and password) and electronic contracts. The encryption method is called Elliptic Curve Cryptography resulting in a project named MediBChain. The drawbacks observed in this model is implementation costs, system migration and interoperability.

Healthcare Industry is witnessing one of the highest growths in revenues and volume of data by EHR/EMR. The is a growing need to store and share data among stakeholders for effective healthcare delivery in a secured manner. While the data is moving towards the cloud, there is a pressing need for its security which can be addressed by blockchain architecture. Few related works are described below.

In this [23] model, the framework enabled medical scientists, Government health authorities and healthcare stakeholders as end-users with data input from physician offices, hospital servers along with existing data storage infrastructure. Permission management is implemented with the decentralization of medical records by cryptography. Permission management firstly involves taking patients consent. The model works on smart contracts with a Proof-of-Work (POW) algorithm.

In this [32] model keeping the heterogeneous nature of healthcare data, a framework has been developed using blockchain and cloud technologies. Moving away from the centralization of data, the data is stored in a distributed environment ensuring better interoperability. Cryptographic techniques for user identity check is a pre-step before adding a new block on the existing chain for giving the requested data.

In this [33] model, a Blockchain Based Data Sharing (BBDS) framework is developed in a cloud environment. The users are verified and tracked for data requested in cryptographic techniques in a lightweight blockchain system. The system is divided into user, system management and storage layers. The user layer is for blockchain miners; the system management is for secure cryptographic authentication and storage layer is the data stored on the cloud for future uses. Further research on this work as indicated by the researchers, is communication and the authentication protocols in this framework.

A system in [34], is named MeDShare for healthcare data and uses a four-layer model. The key focus is monitoring the end-user for preventing malicious activities. The layers are user, data query, data provenance and structuring, and existing database infrastructure. The user layer is for actual research end-users, the data query layer is subdivided into querying that manages the data requests and trigger for mediation between blockchain and live environment, the data proverance and structuring layer is subdivided for authentication, processing, smart contracts, smart contracts database. The final data layer is that of the existing data.

This [24] model is developed on health information of cancer patients. The system is a membership service for healthcare data storage, multiple nodes, and application programming interfaces (APIs). The user verification a pre-step before access to data for blockchain miners. The database is stored on two locations local server and the cloud. The local server stores the patient cancer data, and the cloud organized categorical data. The access control is defined by patients with encryption based on symmetric key pair for each patient. The system performance is measured on privacy, security, and scalability.

This [31] model is designed keeping the patient as the end-user who desire to store their data on the blockchain. The blocks are added with the addition of a new patient with a timestamp. The verification of blocks happens cryptographically and contain the hash info of previous blocks. The advantages of the model indicated by the authors are avoiding single point of failure, patients right to access, ease of access to healthcare data and identification of malicious activity.

This model [35] is based on three layers, namely data sender layer, registration unit and private accessible unit. This model ensures the anonymity of patient data in its cryptographic mechanism. The data sender layer is accessed by patients to input all the data and store it cryptographically by checking the authenticity through the data receiver layer. The registration unit ensures the storage of credentials like username, password, and biometric details of the actual end-users. Private accessible unit is the mediator to regulate the flow of data in a secured channel to the end-user.

In this [36] model a mobile application and wearable devices are synchronized in a tree-based and batching approaches to address performance. The system is user-centric and has six entities. The first is the user with a wearable device on him/her that gives input of health-related data on the pulse, walking metrics of speed and distance, sleep duration, etc. The user has full access to his or her data and can share it for treatment purposes. The second the wearable device is the interface to record and store in a timely fashion. The data is synchronized to the cloud on a real-time basis. The third is the healthcare provider giving data for the patient's treatment, and the fourth is the health insurance layer for verification purpose. The fifth is the blockchain layer that stores the data and gets updated real-time. The sixth is the cloud database that consists of transactional data of user queries and requests.

In [37] continuation to addressing the blockchain issues in cryptocurrencies on secure transactions, the authors mapped the solution to healthcare. They based their system on three aspects: authentication, confidentiality, and accountability. The end target users are researchers, public health authorities, and blockchain miners. Cryptographic techniques are used for user verification, monitoring, data sharing etc. The architecture is designed to integrate with local and existing data storage.

In this [38] model, the researchers designed the system to access the patient information from the wearable device in emergency situations with a private key and a public key. The private key is recognized only through biometric signature of patient. The other features or benefits are decentralized architecture, interoperability, user verification, user monitoring. The further study in this model as indicated by the authors are communication and the authentication protocols.

This model [39] is named Ethereum that gives guaranteed stockpiling of data and ensures user (in this case, the patients) security cryptographically in healthcare on a blockchain architecture. The metrics on which this system is tracked are performance and execution results. The blocks are added with a new patient in the chain and timestamped. There is also user monitoring to prevent malicious activities.

This model [40] focused on a multilevel blockchain-based location sharing scheme in mobile health care technologies to overcome the shortcomings of security and protection and give the real location of the patient. The system consists of a data sender layer, data receiver layer, registration unit and private accessible unit. The sender layer takes in all the patients data checks and encrypts it while the receiver layer authenticates it with the system and stores the data. The registration unit stores the user credentials and handles the log-ins. The private accessible unit ensures the flow of information between two parties on a verified channel.

This model [41] is developed for medical insurance domain in the blockchain framework keeping patients, clinics, insurance companies, and the data storage servers as stakeholders. The model built on Ethereum platform with performance evaluation as an information check guarantee is the critical aspect of medical insurance. The model is built on three layers namely the user, secure system management and secured data storage. The user layer is for authenticated blockchain miners, the system management layer is for secure transactions, and the storage layer is a key layer protected from breach as it could impact the insurance revenues and outgo.

This model [42] focused on solving the problem of identical patient data stored in different locations and multiple databases by blockchain architecture. This model eliminates the replication of records and network congestion. The advantages of the model indicated by the authors are avoiding single point of failure, patients right to access, ease of access to healthcare data and detection of malicious activity on data.

The work [43] Keeping in view the benefits EHR/EMR and diagnosis of patients, the authors have developed a model of healthcare system in a data-sharing scheme. The scheme comprises two blockchains (i) private blockchain (ii) consortium blockchain. Implemented in JUICE the private scheme stores the patient health records while the public scheme generates secure indexes for the data records stored. The data is stored centrally with the authentication of user feature for requested data dissemination.

This model [44] proposes a framework of blockchain in healthcare system for remote patients with wearable devices on them for getting the information. In this model, the miner is selected by the patient for the process to start. The key drawbacks noticed in this model is block validation and authentication protocols details in the architecture.

This [45] model proposes a framework for smart contracts in blockchain for remote patient monitoring. The patient is tracked through sensors for data input after permission. The data validation happens through a consensus mechanism with the system acting only if all the permitted users are present. The key drawbacks noticed in this model is block validation and authentication protocols details in the architecture.

Table 1. Statistics of Contemporary Contributions in Blockchain Technology for Healthdomain

Author	Model /Project Name	Architecture/Key Features	Applied Area/Industry	Drawbacks /Further Study
Badr et al. [13]	-	Protocol of Pseudonym Based Encryption with Different Authorities (PBE-DA) with blockchain	E-Health Platform For Patients	-
Agbo et al. [14]	-	Preferred Reporting Items for Systematic Reviews and Meta-Analysis (PRISMA) Guidelines and A Systematic Mapping Study Process	Highlights The State-of-The-Art In The Development of Blockchain Applications For Healthcare	-
Vazirani et al. [15]	-	Reviewers Identified Studies Via Systematic Searches of Databases Including PubMed, MEDLINE, Scopus, EMBASE, ProQuest, and Cochrane Library with a Total of 71 Studies	Discuss Potential Benefits And Limitations Without Evaluation Of The Model Effectiveness. Some Systems Tested On Live Data	-
McGhin et al. [16]	-	Review Of Existing Blockchain Technologies In Smart Contracts, Fraud Detection, And Identity Verification.	Vulnerabilities Like Mining Incentives, Mining Attacks, and Key Management. Unique requirements of healthcare applications.	Potential Research Opportunities

Gordon et al. [17]	-	Interoperability Through Five Mechanisms: (1) Digital Access Rules, (2) Data Aggregation, (3) Data Liquidity, (4) Patient Identity, And (5) Data Immutability	The Transition From Institution-Centric To Patient-Centric Data Sharing.	-
Zhang et al. [19]	-	Contributions to "Shared Nationwide Interoperability Roadmap" from the Office of the National Coordinator for Health Information Technology (ONC)	ONC requirements and their implications; FHIRChain - Fast Healthcare Interoperability Resources (FHIR) standard for shared clinical data; FHIRChain-based decentralised app for remote cancer care; Key takeaways	-
Kuo et al. [20]	-	Blockchain Technologies, Benefits, Pitfalls, And The Latest Applications	Biomedical And Health Care Informatics Researchers	-
MIT [26]	MedRec&MedRec 2.0	Data Fragmentation, Interoperability, Patient Focus, and Medical Diagnosis Data	Patient Contracts and Linking of Patient And Caretaker	End-point security
Medical Chain [27]	Medical Chain	Hyper Ledger Fabric for access control and MedTokens	Marketplace For Negotiation With Third Parties For Data Access	Non-purchase of tokens and lack of legal qualification for MedToken security
Patientory [28]	Patientory	HIPAA-compliant data storage systems	Tokens on this platform are used for storage space or third party contract functions	-

Christidis et al. and Bocek et al. [29], [30]	-	Eliminating The Privacy Issue In Using Blockchains In Internet-Of-Things (IoT) Space	Patient/user privacy and pharma supply chain	-
Esposito et al. [31]	-	Blockchain on existing cloud-based medical data with security	EHR/EMR system on the cloud	Complying to the privacy laws and regulations
Omar et al. [35]	MediBChain	Elliptic Curve Cryptography	Pseudonymity (username and password) and electronic contracts	Implementation costs, system migration and interoperability
Azaria et al. [23]	-	Permission Management Is Implemented With The Decentralisation Of Medical Records By Cryptography	Smart contracts with a Proof-of-Work (POW) algorithm	-
Kaur et al. [32]	-	Moving Away From Centralisation Of Data, The Data Is Stored In A Distributed Environment Ensuring Better Interoperability Keeping The Heterogeneous Nature Of Healthcare Data	Cryptographic Techniques for user Identity-Check	-
Xia et al. [33]	-	BlockChain Based Data Sharing (BBDS) framework. The System Is Divided Into User, System Management And Storage Layers	Healthcare Blockchain Miners	Communication and the authentication protocols in this framework
Xia et al. [34]	MeDShare	Monitoring The End-User For Preventing Malicious Activities. The Layers Are User, Data Query, Data Provenance And Structuring, And	Research end-users of Healthcare Data	-

		Existing Database Infrastructure.		
Dubovitskaya et al. [24]	-	The System Is A Membership Service For Healthcare Data Storage, Multiple Nodes, and Application Programming Interfaces (APIs)	Cancer Healthcare Blockchain Miners	-
Liang et al. [36]	-	User-Centric Mobile Application And Wearable Devices Synchronized In A Tree-Based and Batching Approaches	Healthcare Provider and Health Insurance Provider	-
Nakamoto et al. [37]	-	Cryptographic Techniques are Used For User Verification, Monitoring, Data Sharing	Researchers, Public Health Authorities, And Blockchain Miners	-
High et al. [38]	-	Designed The System To Access The Patient Information From The Wearable Device In Emergency Situations With A Private Key And A Public Key	Healthcare Provider	Communication And The Authentication Protocols
Li et al. [39]	Ethereum	Guaranteed Stockpiling Of Data And Ensures User (In This Case The Patients) Security Cryptographically In Healthcare On Blockchain Architecture	Healthcare Blockchain Miners	-
Ji et al. [40]	-	Multilevel Blockchain-Based Location Sharing Scheme In Mobile Health Care Technologies	Healthcare Provider	-

Zhou et al. [41]	Ethereum	Information Check Guarantee As To The Critical Aspect Of Medical Insurance Built On Three Layers Namely The User, Secure System Management And Secured Data Storage	Medical Insurance Domain	-
Fan et al. [42]	-	Solving the problem of identical patient data. Eliminates the replication of records and network congestion.	Healthcare Blockchain Miners	-
Zhang and Lin [43]	JUICE	The Private Scheme Stores The Patient Health Records While The Public Scheme Generates Secure Indexes For The Data Records Stored	Healthcare Blockchain Miners	-
Uddin et al. [44]	-	A Framework Of Blockchain In Healthcare System For Remote Patients With Wearable Devices	Patient Selected Healthcare Blockchain Miners	Block Validation And Authentication Protocols Details In The Architecture
Griggs et al. [45]	-	Framework For Smart Contracts In Blockchain For Remote Patient Monitoring	All User Consensus-Based Healthcare Blockchain Miners	Block Validation And Authentication Protocols Details In The Architecture

6. Gap Analysis

In overall, it is imperative that there are profound ways in which the system can be more robust and development oriented. However, there are certain conditions in the system that must be more effectively planned in terms of implementation. While it is clear from the blockchain implementation in other verticals, that if planned for effective implementation, the solution can be resourceful. Still, the gaps like the cost of infrastructure, and the management of the peer-to-peer network holds some significant issues[6].

For instance, in the operational circuit of the blockchain, once a new block hash is created, it is shared among all the peer-to-peer network members for verification. Post the verification and consensus from the peer members, the block is added to the chain successfully. The constraint to be observed is about approval from the number of peer members to approve the transaction. If there are some fictitious nodes in the peer network, then the scope of the fictitious data blocks added to the chain is possible.

Secondarily, the other gap is about transparency in the data view. Though the block might not permit for any change once successfully verified, still the anomaly detection to the system is possible, wherein the peer network independent nodes have tampered.

The case instances that are taken up for the review towards understanding the gaps in the implementation of the blockchain solutions in healthcare domain are more pertinent to the domain, in addition to some of the common problems that are envisaged in the conditions for the usual implementation of the blockchain models. One critical area of the blockchain-related solution implementation conditions is about the IoT devices used to the blockchain system.

IoT based bodily sensors that work as the critical factor that supports the data transmission to the systems if treated as nodes are integral to the blockchain structure. Numerous studies have focused on how the IoT bodily sensors should be more enduring and the ones that are prone to the risk of compromising. If the node level body sensors are prone to malicious attacks, there are potential chances that the data transmitted to the block is a manipulated information, which might impact the conditions of accuracy.

The blockchain system does not rely on any pre-verification process when the record is reaching the block. Only when the record is created to a block, the verification process is initiated. In simple terms, the prevention of tampering to the existing data is possible with the blockchain system, but the scope of manipulated information reaching the block for a record is potentially high.

Such conditions eventually impact the facets on two dimensions. One the time consumed by the distributed nodes that are integral to the system, towards verifying the record hash. The second in terms of seemingly manipulated information entering the block at the genesis record level.

In a scenario of the context mentioned above, when a body sensor implanted into a human body is infected with malware attack, there are potential chances that the data that is transmitted by the system might also have a significant impact if the block hash is created once based on the manipulated data. Considering the enormity and sensitivity related to the data, it is imperative that the security factors of the nodes that are integral to IoT must be taken into account.

The secondary scenario is about the enormity of data that is generated from the implanted devices to the server network. When there are multiple sets of records that are developed for the healthcare system from the implanted devices and the external set of information systems (electronic health records, diagnostic reports, insurance systems etc.) such data must be a comprehensive network sharing. For instance, when the implanted devices to transmit information for the system, it is integral to the blockchain distribution ledger. But the data related to insurance claim verification factually might not require verifying the record, and it doesn't require the information in its implant

IOP Conf. Series: Materials Science and Engineering 1110 (2021) 012001 doi:10.1088/1757-899X/1110/1/012001

device memory. Hence there is a need for a more structured model of data formulation wherein the blockchain-based system can be more significant for authentication of information access to the distinct set of information systems integral to the process.

The other dimension is when the electronic health record of a patient has some vital parameter related information of the patient; such inputs are presumably accessible to hospitals, insurance companies and the patient. If the implanted device in the body provides information to the system, then, there is a need for all the nodes in the block to take in to account the structure, and there is an integral need for a more robust structure of authentication.

It is highly important that the factors discussed above are given due attention, and there is an appropriate kind of system integration to overcome some of the key factors discussed as gaps in the system implementation. If such metrics are developed, then the implementation of the system in the healthcare domain can be more robust and can provide significant outcome[9].

7. Conclusion

Information systems security has significant importance as any information breach might be a very challenging issue for many stakeholders associated with the system. Blockchain, overall, can be defined as a system that ensures the transaction data is secured and monitored for amendments or trans positioning of data from one level to the other. Such system shall help the system in having more secured levels of development. It is imperative from the literature that the blockchain is undeniably high levels of ingenious development. Today, across the verticals, the role of blockchain solutions is becoming an impeccable development. In this report, the emphasis was about the blockchain implementation in the healthcare domain. One of the common factors that are discussed in the literature is about the scope for implementing the system effectively, as the data in healthcare is highly sensitive and the vulnerabilities of blockchain system that must be addressed. If the objectives defined in the system could be researched and the necessary outcomes are developed as solutions, implementation of blockchain in healthcare can be more sustainable.

8. References

[1] Underwood, Sarah. "*Blockchain beyond bitcoin.*" Communications of the ACM 59, no. 11 (2016): 15-17.

[2] Crosby, Michael, PradanPattanayak, Sanjeev Verma, and Vignesh Kalyanaraman. *"Blockchain technology: Beyond bitcoin."* Applied Innovation 2, no. 6-10 (2016): 71.

[3] Wang, Huaiqing, Kun Chen, and Dongming Xu. *"A maturity model for blockchain adoption."* Financial Innovation 2, no. 1 (2016): 12.

[4] Liu, W., S. S. Zhu, T. Mundie, and U. Krieger. *"Advanced block-chain architecture for e-health systems."* In 2017 IEEE 19th International Conference on e-Health Networking, Applications and Services (Healthcom), pp. 1-6. IEEE, 2017.

[5] Mangesius, Patrick, Johannes Bachmann, Thomas Healy, SamrendSaboor, and Thomas Schabetsberger. *"Blockchains in IHE-Based Networks."* Studies in health technology and informatics 251 (2018): 27-30.

[6] XXu, Z., J. Kim, S. Ryu, D. Alevtina, and F. Wang. *"secure Electronic Medical Records Sharing Using Block chain: su-i-gpd-j-100."* Medical Physics 44, no. 6 (2017): 2822.
[7] Sadiku, Matthew NO, Kelechi G. Eze, and Sarhan M. Musa. *"Block chain Technology in Healthcare."* International Journal of Advances in Scientific Research and Engineering 4 (2018).
[8] M. Shilpa, E. Rahul Sharma, and M. Sartaj Singh, *"Big data analytic on block chain across healthcare sector,"* Int. J. Eng. Technol., 2018.
[9] Siyal, Asad, Aisha Junejo, Muhammad Zawish, Kainat Ahmed, Aiman Khalil, and Georgia Soursou. *"Applications of Blockchain Technology in Medicine and Healthcare: Challenges and Future Perspectives."* Cryptography 3, no. 1 (2019): 3.
[10] Gordon, William J., and Christian Catalini. *"Blockchain technology for healthcare: facilitating the transition to patient-driven interoperability."* Computational and structural biotechnology journal 16 (2018): 224-230.
[11] Cocco, Luisanna, Andrea Pinna, and Michele Marchesi. *"Banking on blockchain: Costs savings thanks to the blockchain technology."* Future Internet 9, no. 3 (2017): 25.
[12] Yadav, Vinay Surendra, and A. R. Singh. *"A Systematic Literature Review of Blockchain Technology in Agriculture."* 2017.
[13] Badr, Shaimaa, Ibrahim Gomaa, and Emad Abd-Elrahman. *"Multi-tier blockchain framework for IoT-EHRs systems."* Procedia Computer Science 141 (2018): 159-166.
[14] Agbo, Cornelius C., Qusay H. Mahmoud, and J. Mikael Eklund. *"Blockchain technology in healthcare: a systematic review."* In Healthcare, vol. 7, no. 2, p. 56. Multidisciplinary Digital Publishing Institute, 2019.
[15] Vazirani, Anuraag A., OdhranO'Donoghue, David Brindley, and Edward Meinert. *"Implementing Blockchains for Efficient Health Care: Systematic Review."* Journal of medical Internet research 21, no. 2 (2019): e12439.
[16] McGhin, Thomas, Kim-Kwang Raymond Choo, Charles Zhechao Liu, and Debiao He. *"Blockchain in healthcare applications: Research challenges and opportunities."* Journal of Network and Computer Applications (2019).
[17] Hölbl, Marko, Marko Kompara, Aida Kamišalić, and Lili NemecZlatolas. *"A systematic review of the use of blockchain in healthcare."* Symmetry 10, no. 10 (2018): 470.
[18] Abujamra, Ramzi, and David Randall. *"Blockchain applications in healthcare and the opportunities and the advancements due to the new information technology framework."* Role of Blockchain Technology in IoT Applications 115 (2019): 141.
[19] Zhang, Peng, Jules White, Douglas C. Schmidt, Gunther Lenz, and S. Trent Rosenbloom. *"FHIRChain: applying blockchain to securely and scalably share clinical data."* Computational and structural biotechnology journal 16 (2018): 267-278.
[20] Kuo, Tsung-Ting, Hyeon-Eui Kim, and Lucila Ohno-Machado. *"Blockchain distributed ledger technologies for biomedical and health care applications."* Journal of the American Medical Informatics Association 24, no. 6 (2017): 1211-1220.
[21] Salari, Pooneh, and Bagher Larijani. *"Ethical Issues Surrounding Personalized Medicine: A Literature Review."* Acta MedicaIranica 55, no. 3 (2017): 209.
[22] Cruz-Correia, Ricardo, Duarte Ferreira, Gustavo Bacelar, Pedro Marques, and Priscila Maranhão. *"Personalised medicine challenges: quality of data."* International Journal of Data Science and Analytics 6, no. 3 (2018): 251-259.

[23] Azaria, Asaph, Ariel Ekblaw, Thiago Vieira, and Andrew Lippman. *"Medrec: Using blockchain for medical data access and permission management."* In 2016 2nd International Conference on Open and Big Data (OBD), pp. 25-30. IEEE, 2016.

[24] Dubovitskaya, Alevtina, Zhigang Xu, Samuel Ryu, Michael Schumacher, and Fusheng Wang. *"Secure and trustable electronic medical records sharing using blockchain."* In AMIA Annual Symposium Proceedings, vol. 2017, p. 650. American Medical Informatics Association, 2017.

[25] Radanović, Igor, and Robert Likić. *"Opportunities for use of blockchain technology in medicine."* Applied health economics and health policy 16, no. 5 (2018): 583-590.

[26] MedRec 2.0. Available online: https://medrec.media.mit.edu/technical/ (accessed on. 4 April 2019).

[27] Medicalchain. Available online: https://medicalchain.com/Medicalchain-Whitepaper-EN.pdf (accessed on 4 April 2019).

[28] McFarlane, Chrissa, Michael Beer, Jesse Brown, and Nelson Prendergast. *"Patientory: A Healthcare Peer-to-Peer EMR Storage Network v1."* Entrust Inc.: Addison, TX, USA (2017).

[29] Christidis, Konstantinos, and Michael Devetsikiotis. *"Blockchains and smart contracts for the internet of things."* Ieee Access 4 (2016): 2292-2303.

[30] Bocek, Thomas, Bruno B. Rodrigues, Tim Strasser, and Burkhard Stiller. *"Blockchains everywhere-a use-case of blockchains in the pharma supply-chain."* In 2017 IFIP/IEEE Symposium on Integrated Network and Service Management (IM), pp. 772-777. IEEE, 2017.

[31] Esposito, Christian, Alfredo De Santis, Genny Tortora, Henry Chang, and Kim-Kwang Raymond Choo. *"Blockchain: A panacea for healthcare cloud-based data security and privacy?."* IEEE Cloud Computing 5, no. 1 (2018): 31-37.

[32] Kaur, Harleen, M. Afshar Alam, Roshan Jameel, Ashish Kumar Mourya, and Victor Chang. *"A proposed solution and future direction for blockchain-based heterogeneous medicare data in cloud environment."* Journal of medical systems 42, no. 8 (2018): 156.

[33] Xia, Qi, Emmanuel Sifah, AblaSmahi, Sandro Amofa, and Xiaosong Zhang. *"BBDS: Blockchain-based data sharing for electronic medical records in cloud environments."* Information 8, no. 2 (2017): 44.

[34] Xia, Q. I., Emmanuel Boateng Sifah, Kwame Omono Asamoah, Jianbin Gao, Xiaojiang Du, and Mohsen Guizani. *"MeDShare: Trust-less medical data sharing among cloud service providers via blockchain."* IEEE Access 5 (2017): 14757-14767.

[35] Al Omar, Abdullah, Mohammad Shahriar Rahman, Anirban Basu, and ShinsakuKiyomoto. *"Medibchain: A blockchain based privacy preserving platform for healthcare data."* In International conference on security, privacy and anonymity in computation, communication and storage, pp. 534-543. Springer, Cham, 2017.

[36] Liang, Xueping, Juan Zhao, Sachin Shetty, Jihong Liu, and Danyi Li. *"Integrating blockchain for data sharing and collaboration in mobile healthcare applications."* In 2017 IEEE 28th Annual International Symposium on Personal, Indoor, and Mobile Radio Communications (PIMRC), pp. 1-5. IEEE, 2017.

[37] Nakamoto S. Bitcoin: *A Peer-to-Peer Electronic Cash System. 2008.* http://www.bitcoin.org. Accessed February 11, 2019.

[38] High, Donald R., Bruce Walter Wilkinson, Todd Mattingly, Robert Cantrell, JO'Brien V. John, Brian Gerard McHale, and Joseph Jurich. *"Obtaining a medical record stored on a blockchain from a wearable device."* U.S. Patent Application 15/840,589, filed June 14, 2018.

[39] Li, Hongyu, Liehuang Zhu, Meng Shen, Feng Gao, Xiaoling Tao, and Sheng Liu. *"Blockchain-based data preservation system for medical data."* Journal of medical systems 42, no. 8 (2018): 141.

[40] Ji, Yaxian, Junwei Zhang, Jianfeng Ma, Chao Yang, and Xin Yao. *"BMPLS: Blockchain-based multi-level privacy-preserving location sharing scheme for telecare medical information systems."* Journal of medical systems 42, no. 8 (2018): 147.

[41] Zhou, Lijing, Licheng Wang, and Yiru Sun. *"Mistore: a blockchain-based medical insurance storage system."* Journal of medical systems 42, no. 8 (2018): 149.

[42] Fan, Kai, Shangyang Wang, Yanhui Ren, Hui Li, and Yintang Yang. *"Medblock: Efficient and secure medical data sharing via blockchain."* Journal of medical systems 42, no.8 (2018): 136.

[43] Zhang, Aiqing, and Xiaodong Lin. *"Towards secure and privacy-preserving data sharing in e-health systems via consortium blockchain."* Journal of medical systems 42, no.8 (2018): 140.

[44] Uddin, Md Ashraf, Andrew Stranieri, Iqbal Gondal, and Venki Balasubramanian. *"A Patient Agent to Manage Blockchains for Remote Patient Monitoring."* Studies in health technology and informatics 254 (2018): 105-115.

[45] Griggs, Kristen N., OlyaOssipova, Christopher P. Kohlios, Alessandro N. Baccarini, Emily A. Howson, and ThaierHayajneh. *"Healthcare blockchain system using smart contracts for secure automated remote patient monitoring."* Journal of medical systems 42, no. 7 (2018): 130.

[46] H L Gururaj, A Manoj Athreya, Ashwin A Kumar, Abhishek M Holla, S M Nagarajath, V Ravi Kumar "*A New Era of Technology*", Cryptocurrencies and Blockchain Technology Applications, May 2020, Wiley Digital Library https://doi.org/10.1002/9781119621201.ch1

[47] H L Gururaj, A Manoj Athreya, Ashwin A Kumar,S M Nagarajath, V Ravi Kumar *"Adoption of pets in distributed network using blockchain technology",* August 2020, International Journal of Blockchains and Cryptocurrencies, Inderscience Publishers, https://doi.org/10.1504/IJBC.2020.108996

IOP Conf. Series: Materials Science and Engineering 1110 (2021) 012002 doi:10.1088/1757-899X/1110/1/012002

Yoga Asana Identification: A Deep Learning Approach

Josvin Jose[1], **Shailesh S**[2,3]
1Carmel College of Engineering and Technology Punnapra , Alappuzha ,Kerala , India

2Sacred Heart College, Thevara, Kochi, Kerala, India

3Department of Computer Applications, Cochin University of Science and Technology, Kalamserry, Kochi, Kerala, India

E-mail: *josvinvalekalam912@gmail.com, shaileshsivan@gmail.com

Abstract. Yoga is a healthy practice that originated from India, to rejuvenate a man in his physical, mental, and spiritual wellness. Moving with the brisk technology advancements, there is a vast opportunity for computational probing in all social domains. But still, the utilization of artificial intelligence and machine learning techniques for applying to an interdisciplinary domain like yoga is quite challenging. In this work, a system that recognizes a yoga posture from an image or a frame of a video has been developed with the help of deep learning techniques like convolutional neural networks (CNN) and transfer learning. We have considered images of 10 different asanas for training the model as well as evaluating the prediction accuracy. The prediction model backed with transfer learning shows promising results with 85% prediction accuracy and this system can be considered as an initial step to build an automated yoga image and video analysis tool.

1.Introduction

The innovations in technology and science are happening in a drastic phase, which makes human life more and more hassle-free. Nowadays everybody is aware of its relevance in day to day life. As in every domain, the influence of computers and computer-powered technologies are well established in health care and related domain. Apart from the usual medical practices, other practices like Yoga, Zumba, martial arts, etc are also widely accepted among society as a way to achieve good health. Yoga [1] [2] is a set of practices sprout out in ancient India which deals with the wellness of physical, mental, and spiritual condition of a man. Yoga has gained a big significance in the medical community. The benefits of yoga are improved health, mental strength, weight loss, etc. But Yoga must be practiced under the supervision of an experienced practitioner since any incorrect or bad posture can lead to health problems such as ankle sprain, stiff neck,

IOP Conf. Series: Materials Science and Engineering 1110 (2021) 012002 doi:10.1088/1757-899X/1110/1/012002

muscle pulls, etc. The yoga instructor must correct the individual postures during training. After proper training only, one can practice Yoga on their own. However, in today's situation, peoples are more comfortable in their homes and everyone changes almost everything to online mode. This situation demands a technology-driven Yoga practice [3] [4]. This can be done with the help of mobile applications or with virtual tutoring applications. In both cases, there are vast opportunities to explore computationally for making them more powerful, intelligent, and efficient. We put

forward a classification model for recognizing and identifying a yoga asana from an image or from a frame of a video in this work. The classification model [5] is developed using the deep learning techniques [6] backed with image processing and computer vision methods. The entire work focuses on classifying 10 classes of yoga asanas. The postures like Bridge, Child's, Downward dog, Mountain, Plank, seated forward bend, Tree, Triangle pose, Warrior1, Warrior2 were analyzed and recognized by the deep learning model and then used to identify the exact class of the yoga posture Figure 1.1 shows the sample images of these 10 classes of yoga asana:

Figure 1.1: Yoga asanas

2.Literature Review

Since the domain of interest of this work, Yoga is not much explored, only a few works were found related to this. So, this section is logically partitioned into two, one which describes the works related to Yoga posture classification, and the other part describes some relevant works done in the general problem of human posture estimation and classification. Sruti Kothari [7] explained a computational method using deep learning, particularly CNN, for classifying the Yoga postures from images. They have considered a dataset containing 1000 images distributed over 6 classes for building the classification model. Nearly 85% accuracy was obtained for this work. Hua- Tsung Chen [8] proposed a yoga posture recognition system that can recognize the yoga posture performed by the trainer. In the first step, he used a kinetic to capture the body map of the user and body contour extraction. Then, the star skeleton which is a fast skeletonization technique by connecting from the centroid to other joint parts is done as the next step. From this technique, accuracy of 96% is acquired. Edwin W.trejo [9] introduced a model for pose correction in yoga. the user will receive real-time instruction about the pose correction made by an expert. for this, they used a recognition algorithm based on the AdaBoost algorithm to create a robust database for estimating 6 yoga poses. Finally, 92% accuracy is obtained.[10] Yoga is a traditional exercise that can bring harmony and peace to both body and mind. But self-learning yoga without the help of a instructor is a hard task. But here a solution is proposed for this, a photo of themselves doing the pose is uploaded then it compares to the pose of the expert and difference in angles of various body joints is calculated.[11] An alternative computationally efficient approach for yoga pose recognition in real world is presented.

A 3D CNN architecture is designed to implement real time recognition of yoga poses, it is a modified version of C3D architecture. For computational efficiency computationally fully connected layers are pruned, and supplementary layers such as the batch normalization and average pooling were introduced.as a result an accuracy of 91.5% is obtained.[12]A dataset consisting of 6 yoga asanas is created. For this work CNN which is used to extract features from key points and LSTM to give temporal prediction is used. As a result a accuracy of 99.04% is obtained. Shailesh s[13][14] proposed a way to annotate dance videos based on foot posture(stanas) in an automatic manner. For this, he used transfer learning. By using transfer learning the features from the images are extracted and a deep neural network is used for image classification. The accuracy obtained was a promising one. Matthias Dantone [15] proposes a unique way to estimate human poses from a 2D image by proposing novel, non-linear joint regression. As a joint regressors, they combined 2 layered random forests. A discriminative independent body part classifier was employed in the first layer, independent body part classifier, and in the second layer, by modeling the interdependence and co-occurrence of the part the joint locations is predicted. Muhammad Usama Islam[16] proposed a method to detect different joint points of a human body and from these points, we calculate the various angle to estimate the poses or asanas and the accuracy of it by Microsoft kinetic. By this method, if a person's angle accuracy falls to below 97% we conclude that the person could not get the pose done. Sadeka Haque [17]proposed a unique way to estimate human postures present in a two-dimensional human exercise image by using CNN, for this they proposed a dataset that contains 2000 images of 5 classes of exercise by conducting various experiments they finally achieved 82.68% accuracy. Sven Kreiss [18] introduced a multi-person pose estimation by using bottom-up strategy with the help of two-methods Part Intensity Field (PIF) and Part Association Field (PAF). The PIF method is used to localize body parts and PAF method is used to associate body parts to each other to form a full human pose, this method is suited for delivery robots and self-driving cars. An accuracy of 96% is obtained from this method. Dushyant Mehta [19] created a technique for 3D motion capture by using an RGB camera. the first step of this technique includes CNN that estimate 2D and 3 D pose features along with identifying all visible joints of the individuals. In this technique, he used a new architecture called SelecSLS Net to improve the information flow without decreasing the accuracy. In the second stage, the pose features of each individual are turned into a complete 3D pose estimate by using a fully connected neural network. In the third stage, to the predicted model a space-time skeletal model is fitted, and for each subject, a full skeletal pose in joint angles is returned. Dongyue Lv [20] created a model to train a golf player to make a perfect swing by capturing and remodel the swing movement in a portable way. To increase the capture accuracy a Dynamic Bayesian Network (DBN) model- based golf swing reconstruction algorithm is proposed, a smart motion reconstruction system for Golf swing (SMRG) is used based on the DBN model and kinetic as capturing device.

3. Methodology

The overall architecture of yoga asana classification is illustrated in the figure 3.1. Each step included the pipeline are explained below.

3.1 Dataset

The dataset for our work is created by Anastasia Marchenkova. In the dataset, there are 10 classes of different yoga poses. They are bridge, children, downward dog, mountain, plank, seated forward bend, tree, triangle pose, warrior1, warrior2. There are around 700 images in the dataset evenly distributed among 10 classes.

3.2 Preprocessing

The structural features of the data in the dataset of yoga asanas are particular for each class of asanas. To make it more compatible for further phases of the deep learning pipeline, the data have to be allied. In the pre-processing stage, we are passing the data through 3 different pre-processing steps as shown in figure. The data are collected from different sources so they will be different in their dimensions. In the first step of preprocessing the data of different dimensions are resized to 100x100 pixels. In the next step, we are using a Gaussian filter to filter out the noises in the data while keeping the edges sharp. And in the third and final step, the Histogram Equalization is used to bring the distribution of the image intensities to a normal fashion and increased contrast of the image is obtained.

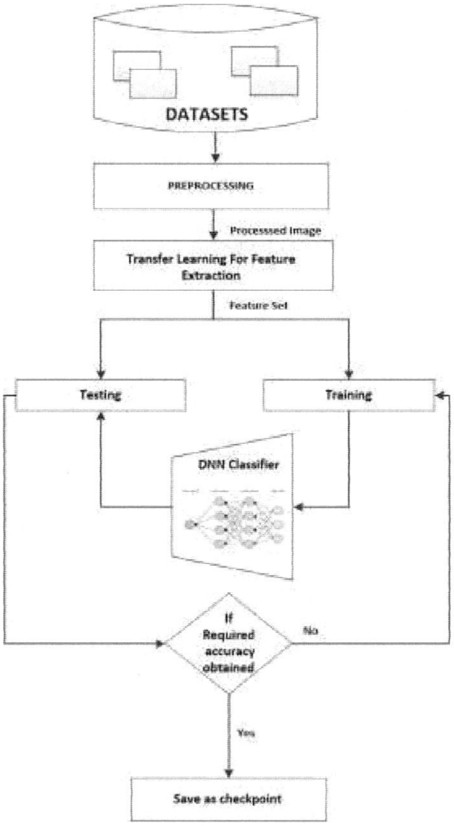

Figure 3.1. Architecture of Yoga Asana Classifier

IOP Conf. Series: Materials Science and Engineering 1110 (2021) 012002 doi:10.1088/1757-899X/1110/1/012002

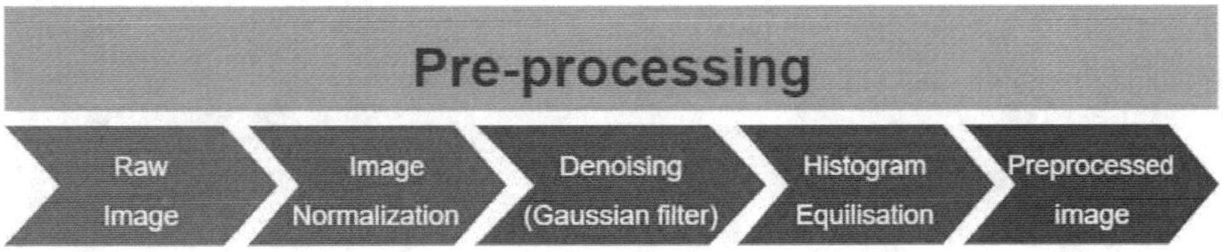

Figure 3.2. Image preprocessing steps

3.3 Transfer Learning (VGG16) for Feature Extraction

Convolutional neural network (CNN) [21] is a popular deep neural network widely used for visual image analysis. The general architecture of CNN is shown below. There are 4 central building blocks on CNN are Convolutional layer, activation layer, pooling layer, and fully connected layer. The Convolutional layer is a filter that passes over a picture and operating some matrix of pixels at a time. The output from this layer will be a layer which speaks about all the pixel the channel observed. The activation layer is a lattice that the convolution layer creates with less estimate than the first picture. This framework is run through an enactment layer, which presents non- linearity to permit the arrangement to prepare itself through back-propagation.

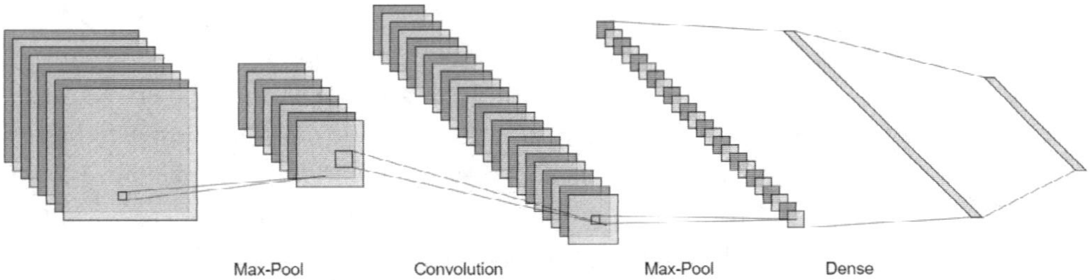

Figure 3.3. Convolutional Neural Network

work is ordinarily (Rectified Linear Unit) ReLu. The pooling layer uses the method "pooling" to encourage down sampling and diminish the measure of the framework. A channel is passed over that comes out of the past layer and chooses one number out of each bunch of values (the most maximum value is selected and this is called max pooling). The fully connected layers are the traditional multi-layer perception structure. Its input is a one-dimensional vector representing the output. The label with the highest probability of acceptance is the classification decision.

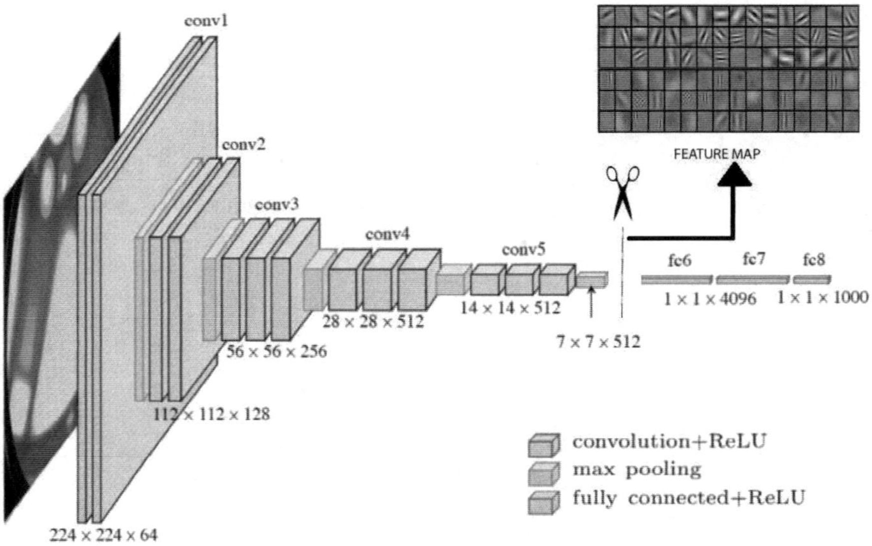

Figure 3.4. Transfer Learning

The Convolutional neuron network was the first attempt to make a CNN model but due to an insufficient amount of data, the performance obtained by the model was not satisfying. To overcome this, instead of manually extracting the features, we tried to obtain the features using transfer learning [22]. Transfer learning involves a method of transferring acquired information to one or more source tasks and is used to enhance the learning of similar target tasks. Although most machine learning algorithms are designed to solve a single task, a subject of ongoing interest in machine learning is to develop algorithms that promote transfer learning. It assumed that fully connected layers collect information relevant to solving a particular problem. For example, the fully connected layers of Alex Net will show the features that are essential for classifying an image into one of 1000 categories of artifacts. For a new problem, The last layer of features of the CNN state pre-trained on ImageNet and use the extracted features to create a new model. In practice, in order to ensure that the system will not fail to learn the previously acquired knowledge, we either maintain a fixed pre-training parameter or adjust it with a small learning rate. The figure below shows a transfer learning model in which a pre-trained deep learning model is used as a feature extractor.

The mathematical definition of transfer learning [23] is given in terms of domain and task. The domain consists of a feature space and a marginal probability distribution P (X) where X = {x1, ..., xn} ∈ χ is a specific domain, = {χ, P (X)} is a task that consists of two components: a label space Y and an objective predictive function f (o) (denoted by $\mathcal{T}$ = {Y, f (o)}. f (o) is acquired from the training data consisting of pairs xi, yi , where xi X and yi Y . The function f(o) can be used to predict the corresponding label while f (x) uses a new instance, x. Given a source domain S and learning task TS , a target domain DT and learning task TT, transfer learning aims to help improve the learning of the target predictive function fT (o) in

DT using the knowledge in DS and TS where DS

DT or TS *f*= TT

3.4 Deep Neural Network Classifier

After the feature extraction process, the next phase is to train a deep neural network [24]. It is a fully connected network with one input layer, eight hidden dense layers, and an output layer. For regularization dropout layers are also added between some pairs of hidden layers. Except for the output layer, the output of all dense layers passes the ReLU activation function. The Softmax function is used in the last dense layer to normalize the network output to the probability distribution on the predicted output classes. Then ADAM optimizer and classification cross- entropy are used as the loss function to compile the network. Th figure x shows the structure of the entire DNN network.

4. Implementation and Result Analysis

	precision	recall	fl-score	support
0	0.60	0.75	0.67	16
1	0.67	0.55	0.60	11
2	0.96	0.89	0.92	27
3	0.98	1.00	0.99	43
4	0.42	0.62	0.50	8
5	0.81	0.92	0.86	24
6	0.80	0.67	0.73	12
7	0.88	0.70	0.78	10
8	0.60	0.50	0.55	12
9	0.88	0.79	0.83	19
Accuracy			0.81	182
macro avg	0.76	0.74	0.74	182
weighted avg	0.82	0.81	0.81	182

As a deep learning application, the implementation of this project was a bit challenging. The work was implemented using python with the help of libraries like TensorFlow, Keras, Panda

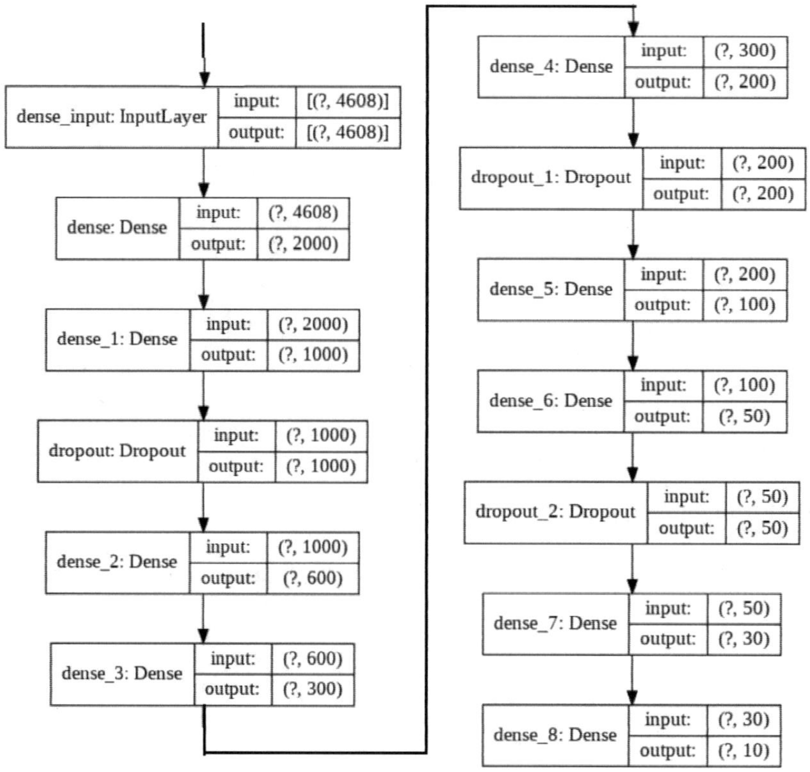

Figure 4.1. Layers of the Architecture

NumPy, OpenCV, PIL. For the transfer learning part, the VGG16 pre-trained model was used with a pre-trained weights magnet. All the images were resized to 100 x 100 and 4608 features were extracted. The entire dataset is split into training and testing sets. The training set is used to build the classifier model, and the test set is used to evaluate the accuracy of the classifier. On testing, we got 85% accuracy the confusion matrix obtained from the actual test labels and predicted test labels were shown in figure the other metrics such as precision, recall, and fi score were also calculated and it is shown in the table 4. Comparing with the traditional CNN architecture and other machine learning models trained with HoG and Hu Moment the proposed transfer learning architecture is more efficient and outperformed others in terms of accuracy and results are given in Table4.

Models	HoG	Hu Moment	CNN	Transfer Learning
Accuracy	70.3%	73.5%	68.3%	85%

Table 1. Accuracy Comparison

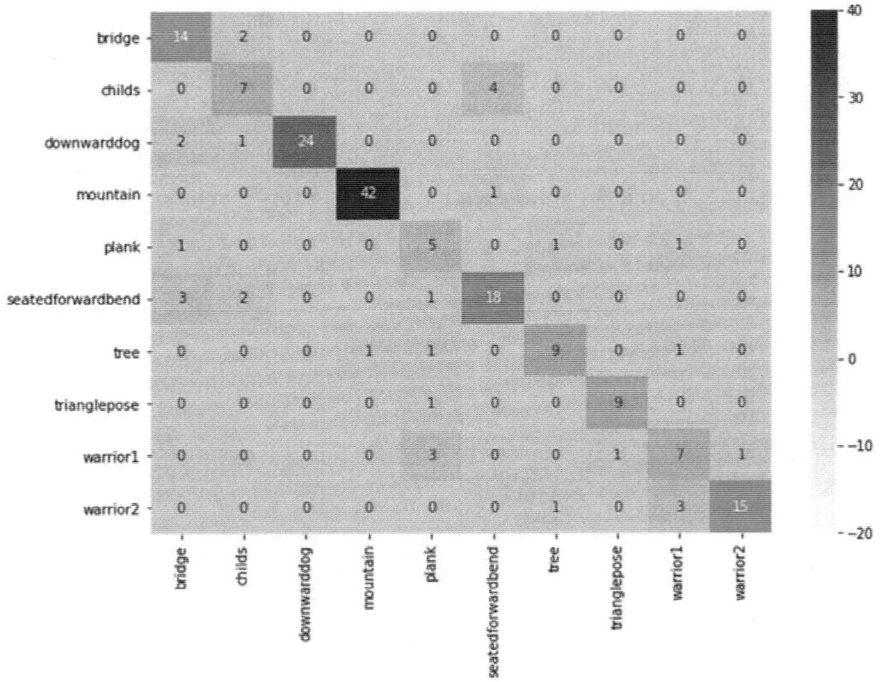

Figure 4.2. Confusion Matrix

5. Conclusion and Future Scope

The advancements and innovations in the field of science and technology pave the path to explore different possibilities in multidisciplinary domains. Newest technologies such as artificial intelligence, machine learning, and computer vision are used to implement many real-time applications in our everyday life. Yoga is one of the widely accepted life practices to nurture the body as well as mind. As a seed to develop a system that can assist humans to practice yoga with a virtual trainer, we have proposed an automatic yoga posture identification system from images or videos. While building this system we have experimented with the state-of-art methods in image classification like CNN, but as the amount of data in the dataset is less those methods did not work well. To get better results we have used transfer learning with VGG16 architecture and pretrained ImageNet weights along with a DNN classifier. The results were quite promising; it gave 82% prediction accuracy. The domain still has numerous possibilities to explore. Apart from the images, video analysis can be done to analyze the movement of the yoga asanas for validating the correctness of the movements. The architectures like 3DCNN, Deep - Pose Estimators, LSTM, GRUs are well suited for video-based analysis.

6. References

[1] Catherine Woodyard 2011 International Journal of Yoga 4 49–54
[2] Ross A, Touchton-Leonard K, Yang L and Wallen G 2016 International Journal of Yoga Therapy ISSN 1531-2054

[3] Giacomucci A 2019 Yoga E-learning platform: Practice with frequency and motivation — UX Case Study
[4] Prasanna Mani,Arunkumar Thangavelu A S and Chaudhari N 2017 International Journal of Intelligent Engineering and Systems 10(3):85-93
[5] Rafi U, Kostrikov I, Gall J and Leibe B 2016 British Machine Vision Conference 2016, BMVC 2016 2016-Septe 109.1–109.11
[6] Yadav S K, Singh A, Gupta A and Raheja J L 2019 Neural Computing and Applications 31 9349–9361 ISSN 14333058
[7] Shruti Kothari 2020 Yoga Pose Classification Using Deep Learning Ph.D. thesis SAN JOSE STATE UNIVERSITY
[8] Chen H T, He Y Z, Chou C L, Lee S Y, Lin B S P and Yu J Y 2013 Electronic Proceedings of the 2013 IEEE International Conference on Multimedia and Expo Workshops, ICMEW 2013
[9] Trejo E W and Yuan P 2018 2018 2nd International Conference on Robotics and Automation Sciences, ICRAS 2018 12–17
[10] Deepak Kumar
[11] Jain S, Rustagi A, Saurav S, Saini R and Singh S 2020 Neural Computing and Applications ISSN 14333058
[12] Yadav S K Yadav, S. K. (n.d.). Real–time Yoga recognition us
[13] Shailesh S and Judy M V 2020 Indian Journal of Computer Science and Engineering 11 89–98 ISSN 22313850
[14] Shailesh S and Judy M V 2020 Intelligent Decision Technologies 14 119–132 ISSN 18758843
[15] Dantone M, Gall J, Leistner C and Van Gool L 2013 Proceedings of the IEEE Computer Society Conference on Computer Vision and Pattern Recognition pp 3041–3048 ISSN 10636919
[16] Islam M U, Mahmud H, Bin Ashraf F, Hossain I and Hasan M K 2018 5th IEEE Region 10 Humanitarian Technology Conference 2017, R10-HTC 2017 2018-Janua 668–673
[17] Haque S, Rabby A S A, Laboni M A, Neehal N and Hossain S A 2019 Communications in Computer and Information Science vol 1035 (Springer Verlag) pp 186–193 ISBN 9789811391804 ISSN 18650937
[18] Kreiss S, Bertoni L and Alahi A 2019 Proceedings of the IEEE Computer Society Conference on Computer Vision and Pattern Recognition vol 2019-June (IEEE Computer Society) pp 11969–11978 ISBN 9781728132938 ISSN 10636919 (Preprint 1903.06593)
[19] Mehta D, Sotnychenko O, Mueller F, Xu W, Elgharib M, Fua P, Seidel H P, Rhodin H, Pons-Moll G and Theobalt C 2019 CVPR (Preprint 1907.00837)
[20] Lv D, Huang Z, Sun L, Yu N and Wu J 2017 Multimedia Tools and Applications 76 1313–1330 ISSN 15737721
[21] Indolia S, Goswami A K, Mishra S P and Asopa P 2018 Procedia Computer Science 132 679–688 ISSN 18770509
[22] Pratt L and Jennings B 1996 Connection Science 8 163–184 ISSN 09540091
[23] Lin Y P and Jung T P 2017 Frontiers in Human Neuroscience 11 ISSN 16625161
[24] Simonyan K and Zisserman A 2015 3rd International Conference on Learning Representations, ICLR 2015 - Conference Track Proceedings

Handwritten Character Recognition of Kannada Language Using Convolutional Neural Networks and Transfer Learning

Parikshith H[1], Naga Rajath S M[2], Shwetha D[3], Sindhu C M[4], Ravi P[5]

[1,2,3,4,5]Department of Computer Science and Engineering, Vidyavardhaka College of Engineering, Mysuru, Karnataka, India

E-mail: parikshithh@gmail.com

Abstract: Recognition of Kannada Handwritten Characters in recent times is one among the active research fields of study. It is one of the challenging topics in the pattern recognition field due vocabulary in large scale, complicated structural hierarchy and different people have different handwriting styles. In this study, we put forward a technique for recognition of characters in kannada which are handwritten based on convolutional neural networks and transfer learning. We have first pre-processed each character to remove noise, cropped each character image and resize the images. After the pre-processing the enhanced pixel values in the image helps in training of the neural network for an efficient classification of characters to their respective classes. We have considered vowels, consonants and numerals of Kannada language which are handwritten. Even though some of the characters in kannada language have similarities in structures our model is capable of classifying the character to their correct respective class. We have also used transfer learning to further improve the model so that any new style of handwritten character can be correctly predicted. Our method is efficient in neural network approach and it achieves competitive performance on the rest of the other traditional recognition methods in the literature.

Keywords: Convolutional Neural Networks, Handwritten Character Recognition, Transfer Learning, Deep learning, Handwritten Kannada Characters

1. Introduction

Handwritten Character Recognition (HCR) is the competency to recognize the character obtained from the image source and interpret the characters. The main challenge is to recognise the character because every person has a different handwriting. Sometimes a person writes a character differently than before. Handwritten texts can be overlapped, and pressure points make differences. There are many ways in which a single character can be written.

The focus of this work is to recognize the handwritten Kannada character image uploaded by users and provide output as a character in digital format. It consists of many stages where the first stage is pre-processing of data which includes removing unwanted spaces and noise from the images, padding and resizing and converting the image to grey scale value. The data set is stored as training and testing data and are fed as input to the neural network as input. Convolutional Neural Network (CNN) model is trained using training data and validation is performed with testing data. Accuracy is improved using Transfer Learning with trained models as input and the output obtained is used for prediction. The UI allows the user to upload the image and the trained model is uploaded in the backend to obtain output in digital format.

2. Related Work

Neural network and TensorFlow techniques are used to Handwritten Character Recognition (HCR). It has a detailed procedure for recognising characters. SoftMax Regression technique is used for classification and character recognition. Diagonal and direction feature extraction methods yielded better results. Feed forward model which is trained by back-propagation algorithm is used. Normalization improved the accuracy of recognition [1].

English characters are recognised and classified using feed forward neural networks. The samples include digits, capital letters, small letters and alphanumeric character. Better results were produced when the model was applied on individual sets rather than all sets combined together [2].

Classification of cursive English script is done by two major phases segmentation and classification. Optical Character Recognition (OCR) which is used to recognize the character and to convert the image format to equivalent machine-readable format. For segmentation the projection methods such as horizontal and vertical projections are used [3].

The approach used to classify Devanagari (Hindi) is Histogram of Oriented Gradients (HOG). The whole process is broken down to segmentation and pre-processing, feature extraction and classification and recognition. Individual characters are classified using ANN. Feature extraction is performed by partitioning the image into six parts. An accuracy of 97.06% is obtained in recognising the Devanagari characters [4].

Characters are recognised using transfer learning and it is recognised even in the poor-quality image of the text. It was done by a pre-trained deep neural network using Inception V3. This model performs better with noisy character images also. It was trained on English Text samples and obtained an accuracy of 90.6% for good quality images and 78% for poor quality images.[5]

Two-stage convolutional neural network is implemented on recognise the Chinese characters which has up to a rotation angle of $\pm 45^\circ$. It was tested on the dataset ICDAR-2013 and obtained an accuracy of 97.38%. It has 7 input channels. The characters are transformed to predict the angle of rotation represented as θ so that the characters are transformed by θ degree to obtain the correct sample. Data augmentation is applied for huge handwriting styles.[6]

Recognition of Kannada characters are performed using two techniques. Tesseract tool has an image with a character in different combinations and sent for pre-processing where the image

IOP Conf. Series: Materials Science and Engineering 1110 (2021) 012003 doi:10.1088/1757-899X/1110/1/012003

is converted to grey scale, denoising technique is used to remove noise and erosion is performed to reduce the thickness of the letter. CNN requires the words to be broken into each character before pre-processing where the counter of letter is formed to obtain x and y coordinates and broken down into each letter and stored in respective folders. Tesseract tool obtained an accuracy of 86% and CNN obtained an accuracy of 87%.[7]

Cursive handwritten text recognition is performed in the language Urdu. The dataset used is UNHD. There are two steps involved in the character recognition namely CNN for feature extraction and bi-directional Long-Short term memory technique for the classification. CNN uses seven layers and is followed by a B-LSTM layer. An accuracy of more than 83% overall is obtained.[8]

Character recognition is performed for Bangla language. The dataset used is Boise State Bangla Handwriting Dataset. It involved segmentation of characters and x and y coordinates marking the length and height of the character is obtained. Features are classified as zonal, pattern and gradient features. Support Vector Machine (SVM) is used for classification purposes. Accuracy is improved by One Versus One (OVO) technique. 96.42% of accuracy is obtained.[9]

Gurmukhi is an Indic language used in Punjab. It has 35 letters and 10 numerals. The dataset was collected from people who wrote the language. The grey scale conversion of the image is done before skeletonization where letters are given proper structure and normalised. The image undergoes segmentation to segregate the letters and the features are extracted. An accuracy of 98.06% is obtained from this.[10]

This paper has compared different techniques used to categorise and recognise Kannada characters and a review is made. The pre-processing is done by converting the image into a binary image, the skews are detected and corrected with an angle of θ. Segmentation is done for a line, word and characters using various techniques. Features are extracted and characters are classified and reviewed using different methods. Accuracy is improved by fusion and classifier methods.[11]

Character classification is done on the basic Kannada characters. This paper has compared the results obtained by two different classifiers namely SVM and K-NN with an accuracy obtained as 93.73% and 91.24% respectively. It is concluded that SVM has higher accuracy for basic characters. It can be further extended for all the Kannada characters. For feature analysis and computing techniques such as Crack code and Fourier descriptors are used.[12]

3. Mathematical Model

The loss function used is the Sparse categorical cross entropy which is actually used in the determination of the performance of the CNN model in this work. The categorical cross entropy formula is given as shown below (1).

IOP Conf. Series: Materials Science and Engineering 1110 (2021) 012003 doi:10.1088/1757-899X/1110/1/012003

$$CCE(p,t) = -\sum_{c=1}^{m} t_{o,c} \; log(p_{o,c}) \quad (1)$$

This categorical cross entropy formula is based on integers, so there is no need for the conversion of targets into categorical format. The formula parameters in this case are 'c' which is to iterate over all classes, m is the total number of classes, $t_{o,c}$ is the binary indicator used to indicate whether c is in the correct class, p_o is the predicted probability. The accuracy is used to evaluate the prediction, it is given by the formula as below (2).

$$Accuracy = \frac{TP + TN}{TP + FN + FP + TN} \quad (2)$$

The accuracy formula consists of a ratio of sum of true negatives and true positives to the sum of true negatives, false positives and false negatives and true positives.

4. Proposed Methodology

4.1 Dataset

The dataset used for this work is the Chars74K dataset. It consists of handwritten kannada character images of all vowels in which there are 16 characters and consonants in which there are 34 characters in total there are 50-character classes. The dataset also consists of handwritten kannada digits from 0 to 9. The dataset consists of images of handwritten characters which have different handwriting styles.

Figure 4.1. Kannada Characters and Numbers

The dataset consisted of around 25 images in each class. This number of images in each class is quite less in number for the developing a model using convolutional neural networks, hence there was a need for more number of images in each character class. The below Figure 4.1 shows all the vowels, consonants and digits in digital form that are present in the dataset.

4.2 Model Architecture

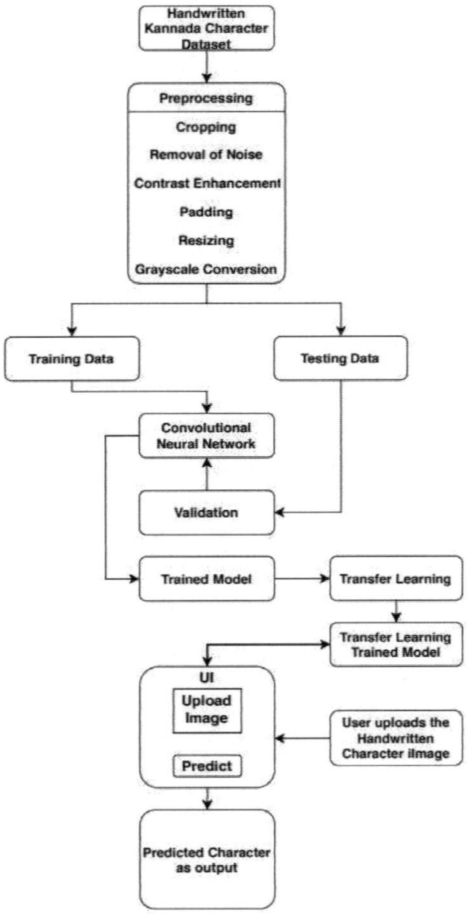

Figure 4.2. Flowchart of System Architecture

This work aims at recognition of handwritten kannada characters precisely and recognised characters are obtained as output in the digital form. The flow of the development of the model involves data pre-processing which involves pre-processing techniques, then the dataset is divided into training and testing data for training and validation purposes. The Figure 4.2 provides the flowchart of system architecture.

After the dataset division in the ratio of 80:20 by shuffling, 80% of data is used for training purposes of the CNN model and rest 20% testing data was used for validation of the trained CNN model. Then the trained model is used for the transfer learning to improve the accuracy so that the model is able to recognize the characters which might have a bit different style of handwriting because transfer learning is used when there is sparse data. Then we have developed a user interface for the user to upload a handwritten character image and get the prediction of the character in the image.

4.3 Data Preprocessing

We have used some of the data pre-processing techniques like cropping of images, removal of noise, contrast enhancement, padding, resizing and gray-scale conversion.

- Cropping - The cropping of images is done to extract a rectangular region of interest around the image, so that extra background spaces are removed from the image and create a bounding box like so only the character is given attention. The cropping is done by taking the difference between the pixel of the image and the background starting pixel.
- Noise Removal - This is used to remove the extra unwanted noise in the image, we have used gaussian blur to remove the unwanted noises.
- Contrast Enhancement - This is used to enhance the brightness between the character image and the background.
- Padding - This is used to introduce new pixels around the edges of the character image, so there will be a uniform boundary around the character for using advanced filtering techniques and it would also help in accuracy as all images that are uploaded for recognition undergoes pre-processing and helps in accurate recognition.
- Resizing - As the deep neural network models expect all the images to be of the same dimensions, we have resized the images to 208x208 pixels.
- Gray-scale Conversion - This is used to convert RGB values which are 24bits into grayscale values which is of 8bits. So this reduces the processing time as there are only 8bits.

One of the samples of the pre-processed image is as shown in the figure 4.3.

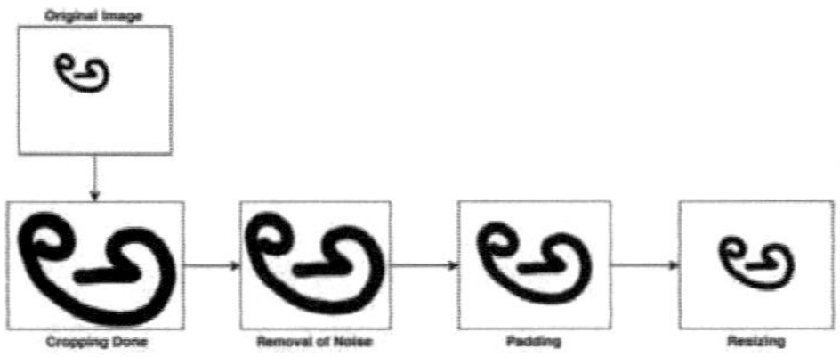

Figure 4.3. Preprocessed sample of a character

4.4 Image Data Augmentation

It is the technique used to artificially increase the dataset when there is sparse image data. This is helpful when only a few data samples are there for each image class. In deep learning, there will be the problem of overfitting due to lack of data samples. So, we also had few image samples in each class so we have used image augmentation to increase the data samples which were around 25 in each class to around 350 per class. This is done using the

ImageDataGenerator function in the keras library. The arguments used in the ImageDataGenerator function are as follows.

Rotation_range: we have used a rotation range of 15^0 so that the images are randomly rotated within this range so we can get different handwriting style rotations.

Height_shift_range: we have used to shift the image pixels in times of height to obtain different dimensions of images so that we obtain the characters in terms of height variability to different handwriting styles.

Shear_range:we have used to control the shear intensity of the images and used a 0.2 as the range for it.

Zoom_range: we have used the zoom range by 0.1 range so that the images are zoomed in this range so that we get images with different dimensional zooms.

Fill_mode: Weh have used fill mode as nearest as our is character specific so it has to be nearest to the original images.

4.5 Implemented Models

The CNN models were implemented to check the accuracy obtained from the machine learning models obtained after the training and validation set is complete. We used the ReLU activation function present in the pytorch functional library to appropriately adjust weights in the neural network and to get correct prediction of results. We first implemented a small convnet with a simple stack consisting of convolution layers which are 3 in number with a ReLU activation and which is followed by a layer of max-pooling. This then has two fully connected layers. In the end we are left with fifty units and hence use softmax activation since this is a multi-class classification. Each convolution block creates a convolution kernel which is convolved with the input layer over a single spatial followed by a nonlinear activation function. The pooling layer downscales the input in both spatial dimensions. The number of filters, filter size and downscale pooling size for 1st, 2nd and 3rd layers are [32, (3,3), (2,2)], [32, (3,3), (2,2)] and [64, (3,3), (2,2)] respectively. After flattening the image i.e converting our 3D feature maps to 1D feature vectors we apply the first fully connected layer which has 64 units and activation function.

After the initial model, we decided to implement a deeper convnet. The stack for each block is as follows - we first applied zero padding i.e., adding rows and columns of zeros at the top, bottom, left and right side of an image. Then convolutions and activation functions are followed by the Batch Normalization layer. This way the activations of the previous layer at each batch are normalized i.e. mean activation is almost 0 and the activation standard deviation close to 1. After this we applied a max-pooling layer. This architecture has 5 such blocks. This is then followed by two fully connected layers. We also used dropout to avoid overfitting. In the end we are left with fifty units and hence use softmax activation since this is a multi-class classification. Kernel size for zero padding, filter size, downscale size in each block is the same and is (1,1), (3,3) and (2,2) respectively. The number of filters for 1st, 2nd and 3rdlayers are 128, 256 and 512 respectively. After flattening the image i.e. converting our 3D feature maps to 1D feature vectors we apply the first fully connected layer which has 64 units and activation function. Finally, we apply relu activation to the last unit to classify into respective class labels. For this CNN model we have achieved the training accuracy of 0.89 and validation accuracy of 0.8692. And for the same model with the same layers of deep network we have trained it for kannada handwritten numerals for which we have achieved a validation accuracy of 0.9670.

The parameters of a neural network are typically the weights of the connections. In this case, during the training stage the learning of the parameters takes place. So, the algorithm itself (and the input data) tunes these parameters. The hyper parameters are typically the learning rate, the batch size or the number of epochs. The model summary is as follows in Table 1.

Table 1. CNN Model summary

Layer (type)	Output Shape	Param
Conv2d-1	[-1,64,96,96]	1,664
Conv2d-2	[-1,128,44,44]	204,928
Conv2d-3	[-1,256,18,18]	819,456
Linear-4	[-1,512]	10,617,344
Linear-5	[-1,256]	131,328
Linear-6	[-1,128]	32,896
Linear-7	[-1,50]	6,450
Total params: 11,814,066		
Trainable params: 11,814,066		

4.6 The Transfer Learning Model

For our transfer learning implementation, we have used the VGG-16 which is a convolutional neural network model. Fully the layers are connected with activation function ReLU, shape = (n_inputs, 256), Dropout with 0.2 chance of dropping, the output is fully connected with log softmax, shape = (256, n_classes). To the model when the extra layers are added, they are set to be trainable by default (require_grad=True). For the VGG-16, the last fully connected layer is only changed. The first fully connected 5 layers and all of the weights that are in the convolutional layers are not trainable. The final outputs from the VGG-16 network are log probabilities for each of the 50 classes in our kannada character dataset.

IOP Conf. Series: Materials Science and Engineering 1110 (2021) 012003 doi:10.1088/1757-899X/1110/1/012003

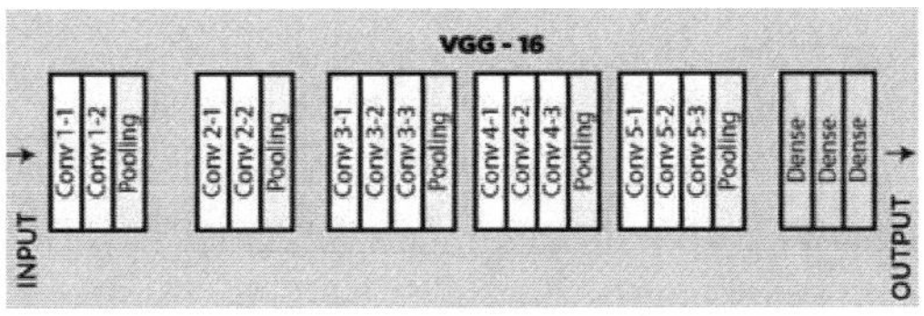

Figure 4.4. VGG-16 Model Architecture

The above Figure 4.4 shows the architecture for the VGG16 model. It is significant to observe that input dimensions for this model were 100x100 instead of 208x208 as shown in the figure. The architecture for this model is as follows: Each block is stacked with convolution layers and at the end of every block a pooling layer to reduce spatial dimension is applied. The output dimensions are increased by a factor of 2 every block. It should also be noted that zero padding is applied before every convolution layer in each block since it is not shown in the figure. The last three fully connected layers were removed as discussed above. Just as a mere experiment, we also decided to implement the VGG16 model architecture from scratch including the fully connected layer to see how it performed. The accuracy after the transfer learning we have achieved is for training we got an accuracy of 0.919 and for validation we have got accuracy of 0.9306.

5. Experiment and Result

We evaluate our models based on the accuracy and loss. We try to minimize the loss function in all the models to achieve better accuracy. Overall, we get a whole simulated model to recognize the handwritten characters. We ran tests for the machine learning model built using ReLU, softmax activation function and the transfer learning model and obtained results for the following measures on validation samples as shown in the table below.

Table 2. Metrics results used in the experiment.

Kannada Characters	**ML Model**	**Accuracy**	**Loss**
Kannada Alphabets (Vowels and Consonants)	CNN Model	0.8692	0.4992
	Transfer Learning Model	0.9306	0.1591
Kannada Numerals	CNN Model	0.9670	0.1374

IOP Conf. Series: Materials Science and Engineering 1110 (2021) 012003 doi:10.1088/1757-899X/1110/1/012003

The graph is plotted for accuracy and loss shown by each model. The accuracy v/s epoch graph has accuracy values on y-axis and epoch values on x-axis. Similarly, for loss v/s epoch graph, loss is plotted on y-axis and epoch on x-axis. It provides an overview of the model behaviour for each epoch run on the model and their accuracy rate.

5.1 Graph Obtained on Training of CNN Model

As shown in Figure 5.1 we can observe that the training accuracy is maintained almost constant between 0.87 and 0.89 with a slight variation. In training loss, we can see a constant decrease.

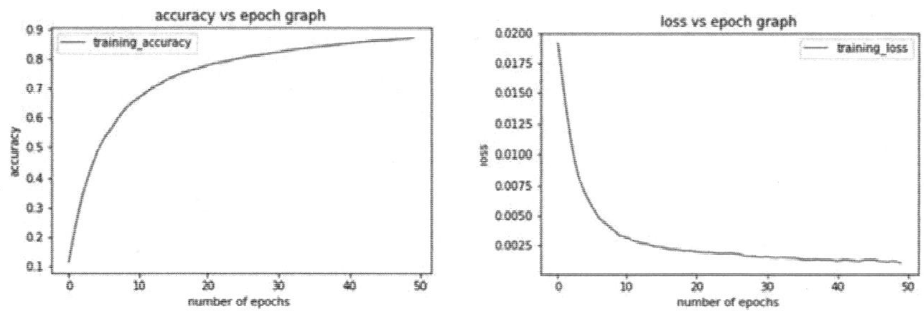

Figure 5.1. Accuracy and Loss Graph of CNN Training

5.2 Graph Obtained on Validation of CNN Model

As shown in Figure 5.2 we can observe that the validation accuracy increases steadily with slight variation in the beginning and we achieve accuracy of 0.8692. The validation loss decreases as the number of epoch's increases but with variations.

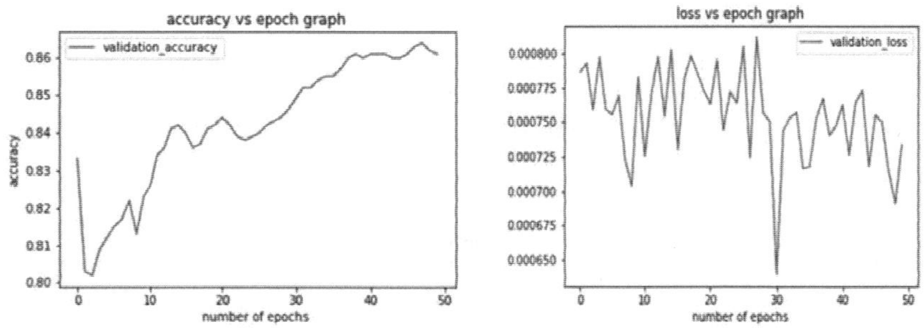

Figure 5.2. Accuracy and Loss Graph of CNN Validation

5.3 Graph Obtained on Training of Pre-trained Model using Transfer Learning

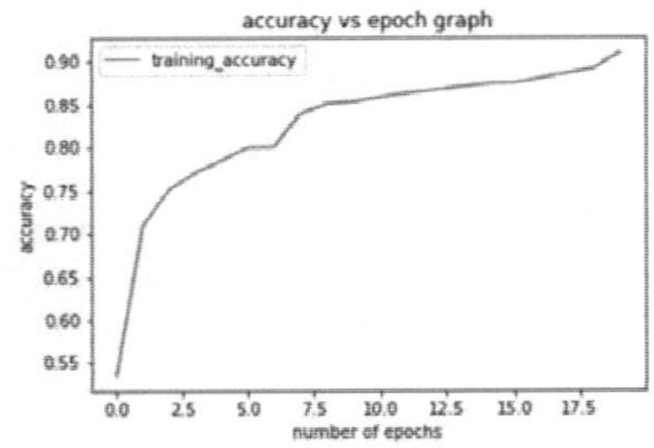

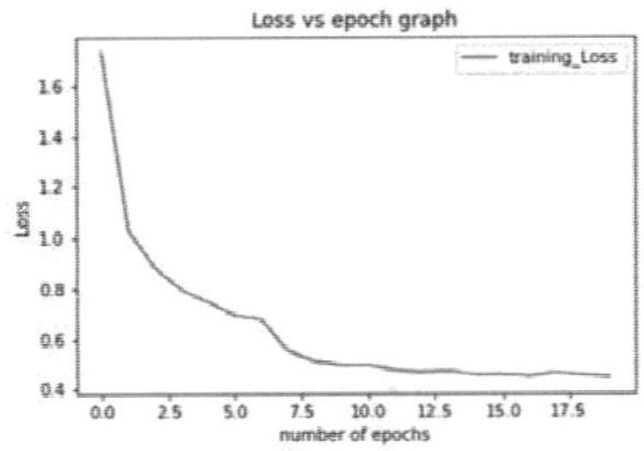

As shown in Figure 5.3 the training accuracy increases and achieves accuracy of 0.9167 for 20 epochs. The training loss decreases, and we can see a constant decrease.

5.4 Graph Obtained on Validation of Pre-trained Model using Transfer Learning

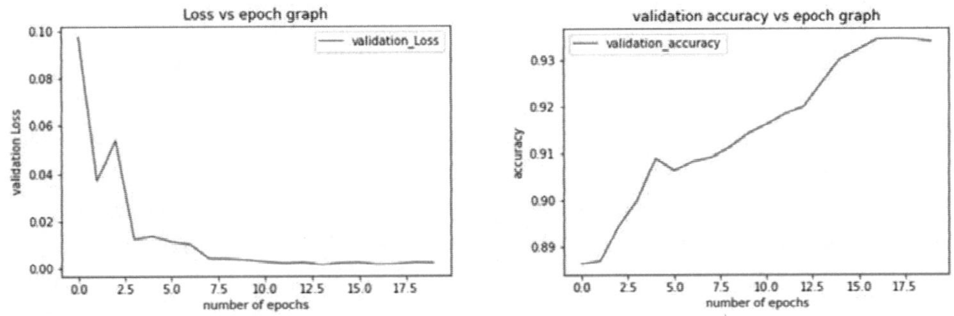

Figure 5.4: Accuracy and Loss Graph of Pre-trained Model Validation

As shown in the above Figure 5.4 the validation accuracy increases at a constant level and achieves an accuracy of 0.9306 for 20 epochs, while the validation loss also decreases as the epochs increase.

We have developed a user interface so that the user can upload the handwritten character image for prediction. The user will get the predicted character output in digital form and also a similarity graph of the character predicted as some characters will have some similarity in the strokes.

The Figure 5.5 below shows the user interface and the predicted character output along with the similarity graph.

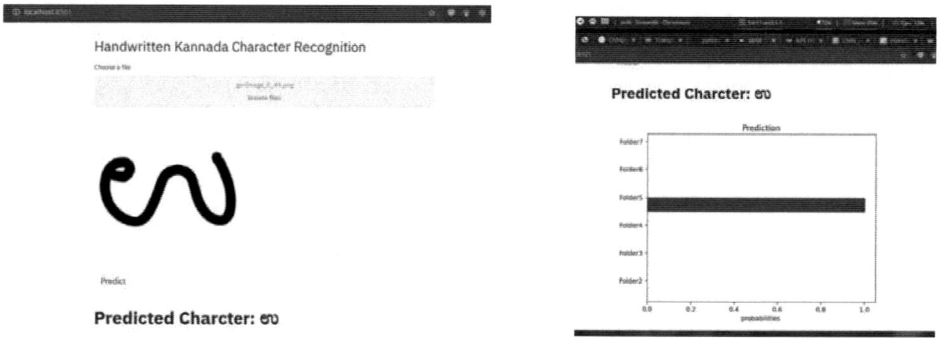

Figure 5.5. User interface of the predicted character along with similarity graph

6. Conclusion

Handwritten kannada character recognition is one of the laborious tasks since there are numerous variations in the styles of writing of every person and also relies on the complication of the languages that are handwritten which needs to be recognized. In this paper we have explored the two main methods used to recognize character from the images consisting of handwritten characters i.e Convolutional Neural Network (CNN) and transfer learning. Before training the CNN model, various pre-processing techniques are applied which includes cropping, padding, denoising and resizing of the image. Since we had a small dataset, the technique called image data augmentation is used to increase the samples in the dataset. Then the augmented images are fed for training CNN model which has both fully connected and convolutional layers and learns the mapping of character image to the respective character label while training. We achieved 86.92% validation accuracy after training the developed CNN model. To develop a good handwritten character recognition system, we had to increase and fine tune the accuracy obtained on training the CNN model. So, we applied a fine-tuning method called transfer learning in which the obtained CNN model is fed and we used VGG16 as the pre-trained network. After fine tuning we achieved a validation accuracy of 93.06% which was much better than the accuracy obtained on training CNN. And we have achieved a validation accuracy of 96.7% for the handwritten kannada numerals recognition.

References

[1] Megha Agarwal, Shalika, Vinam Tomar and Priyanka Gupta, "Handwritten Character Recognition using Neural Network and TensorFlow", *International Journal of Innovative Technology and Exploring Engineering (IJITEE)*, 2019.

[2] Shyla Afroge, Boshir Ahmed and Firoz Mahmud, "Optical character recognition using back propagation neural network", *2016 2nd International Conference on Electrical, Computer & Telecommunication Engineering (ICECTE)*, Rajshahi, Bangladesh, 2016.

[3] Pritam Dhande and Reena Kharat, "Recognition of cursive English handwritten characters", *2017 International Conference on Trends in Electronics and Informatics(ICEI),* Tirunelveli, India, 2017.

[4] Nikita Singh, "An Efficient Approach for Handwritten Devanagari Character Recognition based on Artificial Neural Network", *2018 5th International Conference on Signal Processing and Integrated Networks (SPIN)*, Noida, India, 2018.

[5] Tan Chiang Wei, U.U.Sheikh and Ab Al-Hadi Ab Rahman, "Improved Optical Character Recognition with Deep Neural Network", *2018 IEEE 14th International Colloquium on Signal Processing & Its Applications (CSPA)*, Batu Ferringhi, Malaysia, 2018.

[6] Zhe Li, Lianwen Jin, Songxuan Lai, "Rotation-Free Online Handwritten Chinese Character Recognition using Two-Stage Convolutional Neural Network", *2018 16th International Conference on Frontiers in Handwriting Recognition (ICFHR)*, Niagara Falls, NY, USA, 2018.

[7] Roshan Fernandes, Anisha P Rodrigues, "Kannada Handwritten Script Recognition using Machine Learning Techniques", *2019 IEEE International Conference on Distributed Computing, VLSI, Electrical Circuits and Robotics (DISCOVER)*, Manipal, India, 2019.

[8] Shahbaz Hassan, Ayesha Irfan, Ali Mirza, Imran Siddiqi, "Cursive Handwritten Text Recognition Using Bi-Directional LSTM: A Case Study on Urdu Handwriting", *2019 International Conference on Deep Learning and Machine Learning in Emerging Applications (Deep-ML)*, Istanbul, Turkey, Turkey, 2019.

[9] Nishatul Majid, Elisa H. Barney Smith, "Introducing the Boise State Bangla Handwriting Dataset and an Efficient Offline Recognizer of Isolated Bangla Characters", *2018 16th International Conference on Frontiers in Handwriting Recognition (ICFHR)*, Niagara Falls, NY, USA, 2018.

[10] Manoj Kumar Mahto, Karamjit Bhatia, R. K. Sharma, "Combined horizontal and vertical projection feature extraction technique for Gurmukhi handwritten character recognition", *2015 International Conference on Advances in Computer Engineering and Applications*, Ghaziabad, India, 2015.

IOP Conf. Series: Materials Science and Engineering 1110 (2021) 012003 doi:10.1088/1757-899X/1110/1/012003

[11] K Indira, S Sethu Selvi, "Kannada Character Recognition System: A Review", *InterJRI Science and Technology, Vol. 1, Issue 2*, July 2009.

[12] Ganapatsingh G Rajput, Rajeshwari Horakeri, "Handwritten Kannada Vowel Character Recognition Using Crack Codes and Fourier Descriptors", December 2011.

IOP Conf. Series: Materials Science and Engineering 1110 (2021) 012004 doi:10.1088/1757-899X/1110/1/012004

Predictive Modeling: An Attempt at Predicting Travel Times In Bengaluru Accounting For Geographic And Economic Effects

Jaison Sophia[1], S Althaf[2] and Nambiar Gautham[3]

[1] Student, SMS CUSAT

[2]Assistant professor, NIT Calicut

[3] Student, NIT Calicut

Abstract: The transport system of a country reflects the efficiency and growth of the country. As population increases, the number of vehicles increase, congestion and traffic increase leading to increase travel times, Evolution comes about in the transport system of a country, to increase physical connectivity and economic development, to reduce congestion and travel times. This paper aims to use machine learning algorithm on big data to understand how the effects of rainfall, temperature and pollution help predict travel times. The four machine learning algorithms used include linear regression, ridge regression, random forest regression and elastic net regression. The predicted travel times obtained by all models were compared with the observed travel times in order to determine which model gives better prediction. From the predictive modeling algorithms run on these datasets it is observed that, random forest regression is best suited in predicting travel times in Bengaluru City from i^{th} zone to j^{th} zone in the pth hour of weekdays and weekends after accounting for effects of pollution, temperature, rainfall and economic activity.

1. Introduction

Bengaluru is a city that has relatively higher congestion levels compared to other cities in spite of smaller population. This is due to the limitations of older public transport networks which are primarily road based along with a significant growth in private vehicles [1]. Various reasons can be suggested for congestion including increasing level of motor vehicles, level of infrastructure, psychological factors, policy gaps etc. the average journey speed in Indian cities is also low [3]. Time to travel can vary because of numerous reasons like traffic, poor infrastructure, climatic conditions, mode of travel etc. Travel time variability analysis has been used by policy makers because a high value is allocated by users to the level of services attributes [2].

This paper aims to understand the influence of various factors like precipitation, rainfall, temperature and economic development in uber travel times and tries to develop a predictive model of travel time after considering these factors. These factors were used as a parameter to train the algorithms. For this, quarterly average travel times of two hours ie. 10 am in the morning and 5pm in the evening across weekdays and weekends were considered separately. Quarterly average datasets of Aerosol Optical Depth, precipitation, temperature and nightlight data across wards of Bengaluru are considered. The travel times obtained through the predictive models used in this paper were contrasted against the observed travel times in order to determine which model gives better prediction. The four machine learning algorithms used include linear regression, ridge regression, random forest regression and elastic net regression.

2.Literature Review

Rapid expansion of India's cities has caused the surge in travel demand that in turn has caused a staggering impact on the inadequate transport infrastructure. The escalation in the ownership and use of motor vehicles has led to disturbingly high levels of congestion, noise, air pollution and traffic danger. Mobility and accessibility have declined for segments of the population [4]. Congestion may impede the growth of those cities and have negative aggregate implications for development. Policy makers have used a variety of policies to attempt to address this problem. These policies include increased road provision, the development of new subway lines or bus rapid transit systems, and quantitative restrictions that prevent cars from operating at certain hours depending on their plate number [5]. Denser, more populated cities are slower and there is a hill-shaped relationship between cities per capita income and mobility plus a city's mobility is related to characteristics of its road network. Economic development also brings about better travel infrastructure which facilitates uncongested mobility [6]. Mass rapid transit systems are to be adopted by all cities with populations exceeding one million according to a recommendation made by the National Urban Transport Policy of the Government of India with a view of recognizing the critical need to establish sustainable urban mobility [7]. Accessibility increases total wealth and redistributes wealth and economic opportunities will be lost for those places that do not grow their accessibility as economic opportunities shift to take profit of the accessibility benefits. New infrastructure contributes to agglomeration economies thus increasing aggregate output. Metro rail contributes to greater economic opportunity by improving accessibility especially with regards to the economic growth potential obstructed due to the dense, extensive traffic congestion [8]. Since the Delhi Metro Rail started in 2002, a gradual reduction in vehicle demand and reduced consumption of petrol, diesel and CNG was witnessed, according to 2007 estimates [9].

When Stochastic Response Surface Method was used in a study to investigate the expansion of stochastic response surface of travel time variation under uncertain factors of rainfall intensity and traffic volume, it was found that there is a significant influence of traffic accidents, traffic volume, and amount of rainfall, in travel times on the Hanshin Expressway study area. The most effective indicator in measuring performance of a transportation system was pointed out as travel time distribution. Uncertainties caused by the supply factors, demand factors and other external factors of a transportation system have impacted the properties of travel time distribution. Regression analysis was considered, to formulate a functional relationship between the travel time and various uncertain parameters [10]. This paper controls for precipitation, aerosol optical depth and temperature. According to a study conducted on Korean freeway, increase in rainfall in unit time leads to increased non recurrent traffic congestion and thus reduces travel speed. Average non recurrent congestion surged due to heavy snowfall. Rainfalls and snowfalls occurring from 4:31–8:00 p.m brought about the highest non recurrent traffic congestion. But the rainfall and snowfalls witnessed after 8:00 p.m. or before 7:00 a.m. led to minimum measures in non-recurrent traffic congestion [11]. Direct and indirect effects of atmospheric aerosols and suspended particles have brought about a distressing influence on the human health, visibility, environment and air quality [12]. Statistically significant correlation was found to exist between Aerosol Optical Depth and the RSP (respirable suspended particulates) measured at the air quality monitoring stations. This clearly suggests that the satellite data is reliable source for aerosol-related air pollution studies. Aerosol Optical Depth data covers a more expansive area and has superior spatial resolution when compared with concentrations extracted from the ground-based air quality monitoring networks [13].

Machine learning techniques have been used to predict short term traffic congestion using vehicle trajectories [14]. Change detection in environment can be done through machine learning algorithms specifically classifiers [15]. Random forest is used to work on a road traffic forecasting model because of its exceptional high robustness, performance and practicability [16]. Machine learning methods have also been compared as short term prediction models of travel times [17].

Data

For this analysis travel time data was obtained from Uber movement. Uber provides data by multi-hour period by day, or alternatively, by individual hour by calendar quarter. Uber Movement divides each city into small zones and provides a travel time between zone pairs for each hour of the day [6]. The consolidated data provides quarterly average hourly travel times data across weekends and weekdays between zones from 2016 Q_1 to 2020 Q_2.

Quarterly average data of pollution, temperature and rainfall levels of each zone and intermittent zones were extracted from datasets found in Google Earth Engine. The dataset ERA5 Monthly aggregates - Latest climate reanalysis produced by European Centre for Medium range Weather Forecasts / Copernicus Climate Change Service was used for collecting quarterly average temperature of the 198 Bruhat Bengaluru Mahanagara Palike wards, for the 17 quarters. NASA Earth Exchange Global Daily Downscaled Climate Projections were used to collect quarterly average precipitation across the wards for the 17 quarters. Quarterly average Aerosol optical depth was extracted from the optical depth 047 band of the Moderate Resolution Imaging Spectroradiometer Terra and Aqua combined Multi-angle Implementation of Atmospheric Correction (MAIAC) Land Aerosol Optical Depth gridded Level 2 product produced daily at 1 km resolution (MCD19A2 V6 data product). Nighttime light data are regularly used to measure the intensity of economic activities on the earth surface, and for the estimate socioeconomic parameters. Nighttime light composite data can be extracted from the Visible Infrared Imaging Radiometer Suite (VIIRS) Day-Night Band (DNB) carried by the Suomi National Polar-orbiting Partnership (NPP) Satellite. NPP-VIIRS is equipped with wider radiometric detection range and on-board radiometric calibration, thus giving a more error free nighttime light source for economic modeling [18]. Aerosol Optical Depth, precipitation, temperature and nightlight data which acts as a scale of economic development, are considered as potential attributes affecting the mean travel times of Uber services.

3.Methodology

Supervised machine learning algorithms are those functions that do prediction on past data or out of a sample. They understand the mapping of the function that generates output variable Y from input variable and train it on a portion of the dataset (train dataset) to produce a model which is tested against another portion of the dataset (test dataset). Linear regression is one of the simplest linear methods for regression. Linear regression does not involve any tuning. Ridge regression is used to control complexity, by regularization which involves considering near zero coefficients. L2 regularisation ensures that the coefficients are shrunk by the same factor evicting the possibility of elimination of coefficients and chances of creating sparse models. In ridge regression, high values of alpha restrict the coefficients, thus decreasing generalization. Elastic net is a regularised linear regression combining penalties of both lasso and ridge. The methodology is to record two statistics for each model trained- the R squared value and the Mean squared error, followed by predicting and evaluating the model on the train dataset and test dataset consecutively. R squared value depicts the proportion of variance in the dependent variable/output variable that can be explained by the independent variable/input variable in the regression model. Mean squared error is the average of the square of the difference between the predicted values and the actual value (this difference is termed as error).

The Y dataset contains past travel time data and the X dataset contains the aerosol optical depth measures, nightlight measures, temperature and precipitation measures of source and destination wards separately. To ensure the accuracy of the models, the dataset is split into train and test dataset with the training dataset as 60% of random observations of the original data and 40% constituting the test dataset. After training the algorithms to each train dataset, R squared value, Mean squared error and root mean squared error are calculated following which the same is done in the test dataset too. Travel times are predicted after considering the influence of pollution, precipitation, temperature and economic development, in the source and destination areas of travel, across both the training sets and test sets. Predicted travel times are then plotted against the test data values. Initially linear regression is performed, and values predicted on test dataset, but the results were not satisfactory with R squared value below 1. Linear regression can be described as a classical parametric method that sometimes requires clear cut modeling of nonlinearities and interactions. In a scenario, where the number of observations ‘n’ is greater than the number of variables, linear regression breaks down but this can be easily remedied by shrinkage methods like ridge regression, the lasso or the elastic net and so linear regression is treated as plausibly robust. As a result we run ridge regression on the datasets. For ridge regression, alpha value taken is 4. Elastic net regression algorithm is also trained and used to predict on the test dataset. Unlike the other machine learning algorithms discussed above random forests can be described as nonparametric and give way to the learning of interactions and nonlinearities without having to explicitly model them. They also work better when data mining through data in scenarios having an obviously larger number of observations than the number of variables p or vice verse [19]. Random forests address the overfitting of training data, the n_estimators i.e. the number of trees to build is set as 300, with random state = 0.

The R squared values and the mean squared error obtained from each machine learning algorithm across four datasets which include travel times at 10 am on a weekday, travel times at 10 am on a weekend, travel times at 5 pm on a weekday and travel times at 5 pm on a weekend are tabulated and examined. Along with random forest regression, feature importance was also calculated to understand which features contributed to the prediction process and on how to avoid overfitting.

4. Results

HOD= 10 WEEKDAY

	R Squared	MSE	R Squared	MSE
OLS	-0.001530863	1412032.392	-0.006546337	1399429.541
Ridge regression	0.004187284	1403970.525	0.000481759	1389658.182
Random forest	0.923284884	108158.6532	0.436946299	782829.3174
Elastic Net regression	0.009840496	1396000.207	0.001172658	1388697.606

HOD= 10 WEEKEND

	R Squared	MSE	R Squared	MSE
OLS	-0.058657531	1083200.873	-0.052503569	1065794.578
Ridge regression	0.005883243	1017163.822	0.005556663	1007001.161
Random forest	0.922610132	79184.03275	0.43422916	572915.3903
Elastic Net regression	0.009894879	1013059.182	0.011396898	1001087.176

HOD= 16 WEEKDAY

	R Squared	MSE	R Squared	MSE
OLS	-5.69E-05	1591008.606	-0.000118611	1609395.352
Ridge regression	0.004062491	1584455	0.003496937	1603577.196
Random forest	0.926584659	116797.7942	0.441573297	898622.7535
Elastic Net regression	0.007492507	1578998.126	0.005265886	1600730.596

IOP Conf. Series: Materials Science and Engineering 1110 (2021) 012004 doi:10.1088/1757-899X/1110/1/012004

	HOD= 16		WEEKEND	
	R Squared	MSE	R Squared	MSE
OLS	-1.04E-09	1283543.05	- 0.001063487	1309748.35
Ridge regression	0.00510323	1276992.833	0.004065103	1303038.323
Random forest	0.925349218	95817.49252	0.467462553	696749.059
Elastic Net regression	0.009655601	1271149.668	0.009279947	1296215.446

According to the R squared value of each model used, linear regression and ridge regression fail at predicting travel times with R squared values falling below 0.05 and for the former the values are negative. Elastic net regression models also have no significant values, but there is a slight improvement in the values thus validating the statement that it is a 'regularized optimization' for linear regression bridging ridge and lasso regression [20]. The best predictor model is the random forest regression that provides 92-93% accuracy across training datasets and 43-46% accuracy on test sets. This gap between the accuracy of the training dataset and test sets implies that further tuning and cross validation is needed, but among the four predictive analytics model used, random forest regression suits the best. The below given graphs that plot the predicted travel times by each model against the actual values, best describe the accuracy of the models.

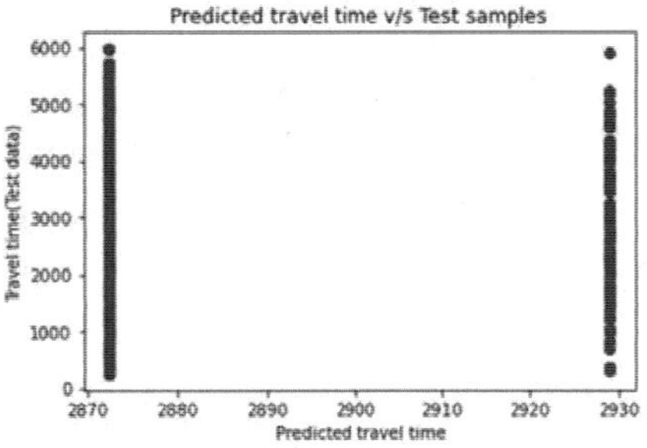

Fig 4.1 Plot of predicted travel times on test dataset against observed travel times using the Linear regression prediction model. It is clear that the travel times predicted were significantly different from observed travel times, thus making the model void for prediction of travel times with the given datasets and variables.

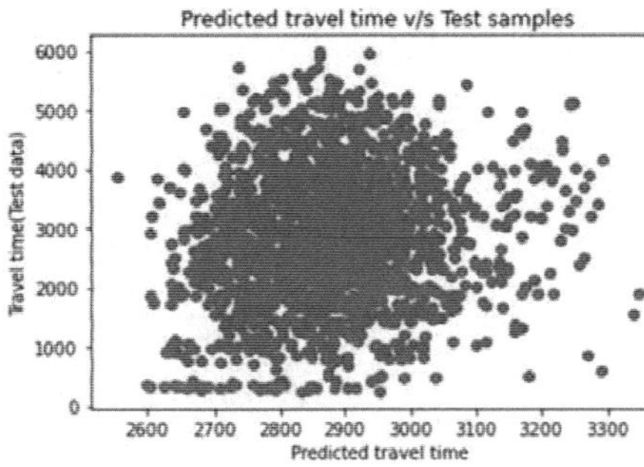

Fig 4.2 Plot of predicted travel times on test dataset against observed travel times using the ridge regression prediction model. It is clear that the travel times predicted were significantly different from observed travel times but it is a comparatively better at prediction compared to linear regression model, But the ridge regression model is void for prediction of travel times with the given datasets and variables.

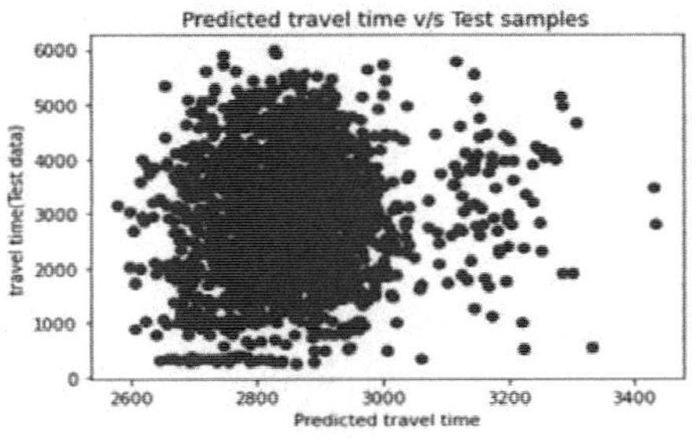

Fig 4.3 Plot of predicted travel times on test dataset against observed travel times using the Elastic Net regression prediction model. It is clear that the travel times predicted were significantly different from observed travel times but an improvement in prediction can be witnessed compared to the above two models, But the the model is still void for prediction of travel times with the given datasets and variables.

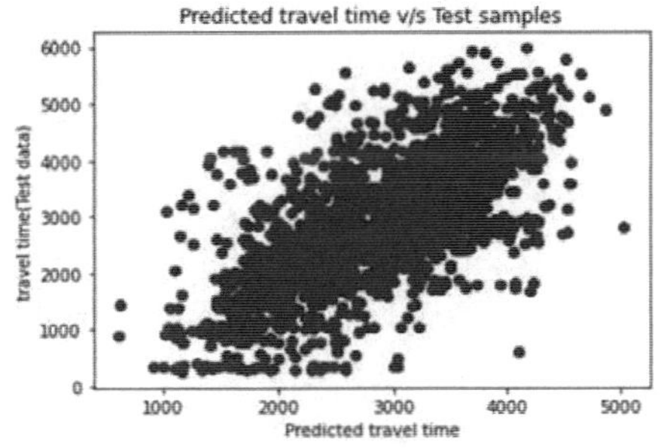

Fig 4.4 Plot of predicted travel times on test dataset against observed travel times using the Random forest regression prediction model. It is clear that the travel times predicted were closer to the observed travel times, thus making the model best ,out of the four models used, for prediction of travel times with the given datasets and variables.

On calculating feature importance of the input variables it was found that economic activity, precipitation and pollution at source has a more significant effect on the travel times compared to other variables used. Better accuracy on the test dataset can be gained by avoiding the other variables. Refer table 1.0

	cols	imp
3	viirs_source	0.343322
1	aod_source	0.299539
0	ppt_source	0.196098
4	ppt_dst	0.041708
5	aod_dst	0.037333
7	viirs_dst	0.035821
2	temp_source	0.029955
6	temp_dst	0.016224

Table 1.0 clearly shows that only three variables have a more significant effect on the prediction model.

5. Conclusion

This paper has identified random forest regression model as an apt model to predict travel times of Bengaluru City using past travel time data and variables like aerosol optical depth, precipitation, temperature and night light data (as a measurement of economic activity). In the event of growing usage of vehicles and increased urbanization this model can help predict the travel times and understand the effect of environmental factors and economic factors on a person's travel time. It can be helpful for new development and construction decisions. It can also influence decisions regarding other high-speed rail and mass rapid transit systems. But the Uber dataset does not strictly imply a person's travel time data. And considering the other modes of transport can also improve the applications of the prediction model. But there is a large difference in the R squared values gained from the train dataset and test dataset. This is because of overfitting and maybe solved by excluding the variables with minimum feature importance and hyperparameter tuning.

References:

[1] Boston Consulting Group 2011 Unlocking cities: the impact of ridesharing across India.

[2] Duran-Hormazabal E & Tirachini A 2016 Estimation of travel time variability for cars, buses, metro and door-to-door public transport trips in Santiago, Chile *Research in Transportation Economics* , **59**

[3] Alam M A & Ahmed F 2013 Urban transport sysatems and congestion: a case study of Indian cities *Transport and Communications Bulletin for Asia and the Pacific*

[4] Pucher J, Korattyswaropam N, Mittala N and Ittyerah N 2005 Urban transport crisis in India. *Transport Policy* **12** pp 185-198

[5] Akbar P A & Duranton G 2017 Measuring the cost of congestion in highly congested city: Bogota

[6] Akbar P A, Couture V, Duranton G & Storeygard A 2018 Mobility and congestion in urban India .

[7] Ministry of Urban Development Government of India 2014 National Urban Transport Policy 2014

[8] Sharma R & Newman P 2018 Does Urban Rail Increase Land Value in Emerging Cities? Value Uplift from Bangalore Metro *Transportation Research Part A: Policy and Practice* **117**

[9] Sreedharan E 2007 Delhi Metro- the changing face of urban public transport in india *Indian Journal of Transport Management* **56**

[10] Chalumuri R S & Yasuo A 2013 Modelling travel time distribution under various uncertainties on Hanshin expressway of Japan *European Transport Research Review* .

[11] Chung Y 2012 Assessment of non-recurrent congestion caused by precipitation using archived weather and traffic flow data *Transport Policy* **19** pp 167-173.

[12] Gunaseelan I, Bhaskar B & Muthuchelian K 2014 The effect of aerosol optical depth on rainfall with reference to meteorology over metro cities in India *Environmental Science and Pollution Research* , **21**

[13] Lau Kai Hon LI C MAO , J & CHEN J.-C. 2003 A new way of using MODIS data to study air pollution over Hong Kong and the Pearl River Delta *Optical Remote Sensing of the Atmosphere and Clouds III;* **4891**

[14] Elfar A, Talebpour A, & Mahmassani H S 2018 Machine learning approach to short-term traffic congestion prediction in a connected environment *Transportation Research Record* **2672** pp 185-195

[15] Chan Jonathan Cheung-Wai, Chan Kwok-Ping & Yeh Anthony Gar-On 2001 Detecting the nature of change in an urban environment: A comparison of machine learning algorithms *Photogrammetric Engineering and Remote Sensing* **67.2** pp 213-226.

[16] Y. Liu and H. Wu 2017 Prediction of Road Traffic Congestion Based on Random Forest *2017 10th International Symposium on Computational Intelligence and Design (ISCID)* pp. 361-364

[17] D. Nikovski, N. Nishiuma, Y. Goto and H. Kumazawa 2005 Univariate short-term prediction of road travel times *Proceedings. 2005 IEEE Intelligent Transportation Systems, 2005* pp. 1074-1079

[18] Li X, Xu H, Chen X, & Li C 2013 Potential of NPP-VIIRS Nighttime Light Imagery for Modeling the Regional Economy of China *Remote Sens* **5 (6)** pp 3057-3081.

[19] Gromping U 2009 Variable Importance Assessment in Regression: Linear Regression versus Random Forest *The American Statistician* **63**

[20] Hans C 2011 Elastic Net Regression Modeling With the Orthant Normal Prior *Journal of the American Statistical Association* **106**

Cascade correlation neural network with deterministic weight modification for predicting stock market price

K VELUSAMY[1] and R AMALRAJ[2]

[1,2]Department of Computer Science
[1,2]Sri Vasavi College, Erode, Tamilnadu, India
[1]kvelusamy85@gmail.com and [2]r_amalraj05@yahoo.co.in

Abstract. A new weight updating approach is proposed for cascade correlation neural network (CCNN) in this paper. The deterministic weight modification (DWM) algorithm is used to adjust the connection weights of CCNN. The introduced new method can improve the global convergence capability of the conventional CCNN and optimally reduced the system error. The proposed DWM+CCNN prediction algorithm is applied well-known stock market dataset in order to evaluate the robustness and efficiency. The experimental results are confirmed that the proposed DWM+CCNN algorithm is achieves higher performance in terms of convergence rate and the capability of global converges.

Keywords: Artificial Neural Networks, Back Propagation Neural Network, Cascade Correlation Neural Network, Deterministic Weight Modification, Stock Index Prediction

1. Introduction

Artificial Neural Networks (ANNs) is well-known information processing system that can approximate any continues function and its universal approximation property makes suitable for nonlinear model [1]. The architecture of the ANNs is an organized by the number of network layers including the input layer, hidden layer, and output layer. An efficient architecture of ANNs is determined by the network layers, connection weights, learning rate, and learning algorithm [2]. However, the number of neurons in both input and output layer is decided according to a given solution. In the hidden layer, deciding the number of neurons is also a crucial and challenging task and another N-P hard problem [3, 4]. If selecting a more number of neurons in the hidden layer, the learning process falls into over fitting. Similarly, if selecting a small number of neurons in the hidden layer, the learning process falls into under fitting [5, 6].

In ANN, the back propagation neural network (BPNN) is well-known neural network architecture because of its straightforwardness and high problem-solving capability. It is supervised and feed forward neural network (FFNN) architecture. However, the BPNN algorithm is to decide its structure before starting the learning process. In this circumstance, there are many drawbacks, including slow convergence rate, failure to find global minima and fall into overtraining or undertraining [5]. Generally, the size of the BPNN architecture is determined by using trial-and-error technique [7]. On the other hand, many research works have been carried out to choose the proper network structure of BPNN [3, 5, 8]. However, the convergence rate of these enhanced BPNN architectures is slow because there is no thumb rule to decide the proper structure [9].

The pruning and constructive neural networks are an alternative way to determine the network sizes that can be automatically found the optimal size of architecture [4, 10]. From these two methods, the pruning algorithm starts with large networks then removing the hidden neuron one by one until the network achieves the optimal error. The constructive neural network algorithm learns begin with zero hidden neuron, then add the hidden neuron one by one until the network is obtaining optimal results. Compared with these two architectures, the constructive neural network is more efficient [11]. The CCNN is a well well-known CNN architecture. It is a FFNN architecture like BPNN. But it is a more powerful architecture compared with BPNN [12]. The weight adjustment of conventional CCNN is done with the help of Quickprop (QPROP) algorithm [2, 12]. The QPROP algorithm is a fast-learning algorithm and uses successive gradient error surface values in weight space to assess the minimum and move toward to global optimal values. However, the gradient based learning algorithms does not produce better generalization ability, fall into local minima and poor convergence rate [13].
In this research work, the DWM algorithm is used to change the weights of the CCNN in order to enhance the performances. The major aim of this research work is to implement an efficient prediction model that can automatically adjust the network size according to a given solution with desirable accuracy. The contributions of proposed research work are as follows:

- The proposed model is enhancing the convergence rate and the prediction accuracy.
- The proposed model is managing the over fitting or under fitting during learning process.
- The efficient prediction model is constructed with the fast-learning process and low memory consumptions.

The paper is structured as follows: In section 2, conventional CCNN architecture is described. The detail of the proposed method is described in section 3. The experimental outcome and discussion are demonstrated in section 4. Finally, the conclusion is presented in the section 5.

2.Cascade Correlation Neural Networks (CCNN)

The CCNN is well-known CNN technique developed by S.E Fahlman and C. Lebiere (1990) [14]. The major goal of the CCNN is automatically increased the size of network structure during the learning process [14, 15]. Two types of the key ideas are in the conventional CCNN. First, cascade structure is leads to create a powerful feature detector. Second, the objective function is used to create and train the new hidden units is a new one. The aim of an objective function is to maximize the degree of correlation between the recent hidden units' output and the remaining network output error.

The CCNN has many advantages compared with non-constructive neural networks, including learning speed is high, own size, depth and topologies are determined dynamically. There is not perform the back propagation of the error and only one-layer weights are trained when hidden neurons is added to the active network. Initially, the learning process of the CCNN is started with minimal network. The input unit is directly linked to the one or more output units through adjustable weight and bias input is +1, but without hidden neurons. At end of the iteration, calculates the error values. If the network error achieves a valid error, then discontinue the learning process. Otherwise, the new hidden unit adds one by one to the active networks up to reach reasonable errors [16].

3.Proposed DWM+CCNN Prediction model

The goal of the weight adjustment is to reduce the network error and improving the architecture performances. In some of the improved CCNN, fast learning algorithms, namely QPROP [14, 16, 17], RPROP [18], and GA [19] are used for its weight adjustments during the learning process for reducing error and convergence speed. Still, the above-mentioned algorithms have natural shortcomings such as poor convergence, local mini-ma. To enhance the performance of the DWM+CCNN prediction algorithm, DWM is applied for adjusting weights of the standard CCNN. The purpose of DWM is to adjust the weights of network in a controlled mode for the period of the learning process. The DWM learning algorithm is used to find the global value, least error signal and very rapid convergence rate. Thus, DWM process is to decrease the error by altering the weights of the CCNN in a deterministic way [20]. The figure 1 shows the overview of proposed DWM+CCNN prediction model.

3.1 Weight adjustment between hidden and output layer

In order to make the outputs so close to the expected outcome, the weight modification is implemented between the hidden and output layer. The specific output o neuron of the output layer of the error is selected above the MSE as follows,

$$E_o(i) = \frac{1}{2} \Sigma_{p=1}^{P} [t_{po} - y_{po}(i)]^2 \geq \frac{E(i)}{o} \quad \textbf{(1)}$$

The objective is to change the weights $w_{ko}(i+1)$ associated to the output o neuron and remaining weights in the network will unaffected. The outcome of the after that iteration will become close to the predictable output. This weight change process can speed up of convergence rate and get away from local minimum. It is noted that o is any output neuron for fulfilling the equation (1). Thus o can be either randomly chosen among all appropriate values fulfilling equation (1) or can be one i^{th} the largest MSE. (i.e $E_o(i) > E_{o'}(i) \forall 0$). The output $y_{ko}(i+1)$ in next iteration wants to be close up to the objective value t_{po} the predictable output is measured as

$$y_{po}(i+1) = y_{po}(i) + \beta(t_{po}(i) - y_{po}(i)) = \beta t_{po}(i) - (1-\beta)O_{po}(i) \quad \textbf{(2)}$$

Where, $0<\beta<1$ for all data points P. To obtain the next iteration weights $W_{ko}(i+1)$ from the above, the output becomes

$$y_{po}(i+1) = f(\Sigma_{k=1}^{K} w_{ko}(i+1)\bar{y}_{pk}(i+1) = f(\Sigma_{k=1}^{K} w_{ko}\bar{y}_{pk}(i+1) + \Delta w_{ko}(i+1)\bar{y}_{pk}(i+1)) \quad (3)$$

$$\Sigma_{k=1}^{k} \Delta w_{ko}(i+1)\bar{y}_{pk}(i+1) = \varepsilon_{po} \quad (4)$$

Where,

$$\varepsilon_{po} = f^{-1}(y_{po}(i+1)) - \Sigma_{k=1}^{K} w_{ko}(i)\bar{y}_{pk}(i+1) \quad (5)$$

$$\varepsilon_{po} = f^{-1} \ln(x/_{1-x}) \quad (6)$$

Therefore, the following matrices to compute $w_{ko}(i+1)$ by considering Equation (3),

$$\begin{bmatrix} \bar{y}_{11}(i+1) & \cdots & \bar{y}_{1k}(i+1) & \cdots & \cdots & \bar{y}_{1K}(i+1) \\ \vdots & \ddots & \vdots & \ddots & \ddots & \vdots \\ \bar{y}_{p1}(i+1) & \cdots & \bar{y}_{pk}(i+1) & \cdots & \cdots & \bar{y}_{pK}(i+1) \\ \vdots & \ddots & \vdots & \ddots & \ddots & \vdots \\ \bar{y}_{P1}(i+1) & \cdots & \bar{y}_{Pk}(i+1) & \cdots & \cdots & \bar{y}_{PK}(i+1) \end{bmatrix} \times \begin{bmatrix} \Delta W_{ko}(i+1) \\ \vdots \\ \Delta W_{ko}(i+1) \\ \vdots \\ \Delta W_{ko}(i+1) \end{bmatrix} = \begin{bmatrix} e_{1o} \\ \vdots \\ e_{po} \\ \vdots \\ e_{Po} \end{bmatrix} \quad (7)$$

The expected output by updating $w_{ko}(i+1)$ with the pseudo inverse obtained least squared error condition [21].

$$W_{ko}(i+1) = \Sigma_{p=1}^{P} e_{po}\bar{y}_{pk}(i+1) \Big/ \Sigma_{p=1}^{P} \bar{y}_{pk}(i+1) \quad (8)$$

both $w_{ko}(i+1)$ and its squared error are calculate for each hidden neurons k as follows,

$$E_o^{(k)}(i+1) = 1/_2 \, \Sigma_{p=1}^{P} \Sigma_{o=1}^{O} [t_{po} - y_{po}(i+1)]^2 \quad (9)$$

The current hidden neurons , is selected to update $w_{ko}(i+1)$ and if its root MSE is the smallest than the previous iteration error, stop the training the network model, Otherwise the training process goes on and new hidden neuron adds to the current networks. i.e $E_o^{(k)}(i+1) \leq E_o^{(k)}(i+1)$ for all k.

3.2 Weight adjustment between the input layer and hidden layer

The idea of the weight changing between input layer and hidden layer is to get away from the problem of local minima by reducing error of network into a targeted value in a deterministic way. The predictable value of the network is defined in the next iteration. The network error of the subsequently iteration will be earlier to the predictable value. The DWM algorithm is fall into local optima at i[th] iteration. An error of the output layer is selected above the MSE

$$E_p(i) = \frac{1}{2} \Sigma_{o=1}^{O} [t_{po} - y_{po}(i+1)]^2 \geq \frac{E(i)}{P} \tag{10}$$

The goal of the weight modification of $W_{ko}(i+1)$ for $n = 1, 2, \ldots N$ is the output of next iteration so closed with predictable output and escaped from local minima. Since the output $y_{po}(i+1)$ of the output layer in subsequently iteration required to be close to the objective value $t_{po}(i+1)$, consider the expected output as

$$E_p(i+1) = \frac{1}{2} \Sigma_{o=1}^{O} [t_{po} - y_{po}(i+1)]^2 \geq \frac{\lambda E(i)}{P} \tag{11}$$

Where, $0<\lambda<1$. The error of each data point is measured by the average values. Thus, for any exacting output neurons O.

$$E_{po}(i+1) = \left| t_{po} - y_{po}(i+1) \right| = \sqrt{\frac{2\lambda E(i)}{P}} \tag{12}$$

It can select randomly or $E_{po}(i) \leq E_{po}(i+1)$ for all o. After choosing

$$O_{po}(i+1) = \begin{cases} 1 - \sqrt{\frac{2\lambda E(i)}{PO}} & if\ t_{po}=1 \\ \sqrt{\frac{2\lambda E(i)}{PO}} & if\ t_{po}=0 \end{cases} \tag{13}$$

We also have,

$$O_{po}(i+1) = f(\Sigma_{k=1}^{K} w_{ko}(i+1)\bar{O}_{pk}(i+1)) = f(\Sigma_{k=1}^{K} w_{ko}\bar{O}_{pk}(i+1) + w_{ko}(i+1)\bar{O}_{pk}(i+1)) \tag{14}$$

Every hidden neuron will not be customized, however only one k hidden neuron will be chosen at a time. Such that $o_{pk}(i)$ is close to the 2-tail area for all k.

$$k.\ \min(\bar{O}_{pk}(i), 1 - \bar{O}_{pk}(i)) \leq \min(\bar{O}_{pk}(i-1), 1 - \bar{O}_{pk}(i-1)) \tag{15}$$

Additionally, while now near to the local minimum, therefore, $\Delta\bar{O}_{pk}(i+1) \approx 0$ for $\forall k \neq k$ with help of these above two simplifications, we have

$$\Delta\bar{O}_{pk}(i+1) = \frac{f^{-1}(\bar{O}_{pk}(i+1) - \Sigma_{k=1}^{K} w_{ko}(i+1)\bar{O}_{pk}(i+1))}{w_{ko}(i+1)} \tag{16}$$

Hence, by using equation, $\bar{o}_{pk}(i+1)$ can be obtained as follows,

$$\bar{O}_{pk}(i+1) = \bar{O}_{pk}(i) + \Delta\bar{O}_{pk}(i+1) \tag{17}$$

Furthermore,

$$\bar{o}_{pk}(i+1) = f(\textstyle\sum_{n=1}^{N} \bar{w}_{nk}(i)x_{pm} = f(\sum_{n=1}^{N} \bar{w}_{nk}(i)x_{pm} + \sum_{n=1}^{N} \bar{w}_{nk}(i+1)x_{pm}) \tag{18}$$

And, thus, $\Delta\bar{w}_{nk}(i+1)$ can be obtained for all as

$$\Delta\bar{w}_{nk}(i+1) = \bar{\varepsilon}_{pk}x_{pn} \Big/ \textstyle\sum_{n=1}^{N} x_{pn}^{2} \tag{19}$$

Where,

$$\bar{\varepsilon}_{pk} = f^{-1}(\bar{o}_{pk}(i+1) - f^{-1}(\bar{o}_{pk}(i)) \tag{20}$$

The process of weight modification is done in two related ways in CCNN using DWM. That is, they both changes of weight are to decrease the total error of the network model. However, the DWM is used to update weight between hidden layer and output layer for decreasing comprehensively by affecting the output of the output layer. In the output neurons

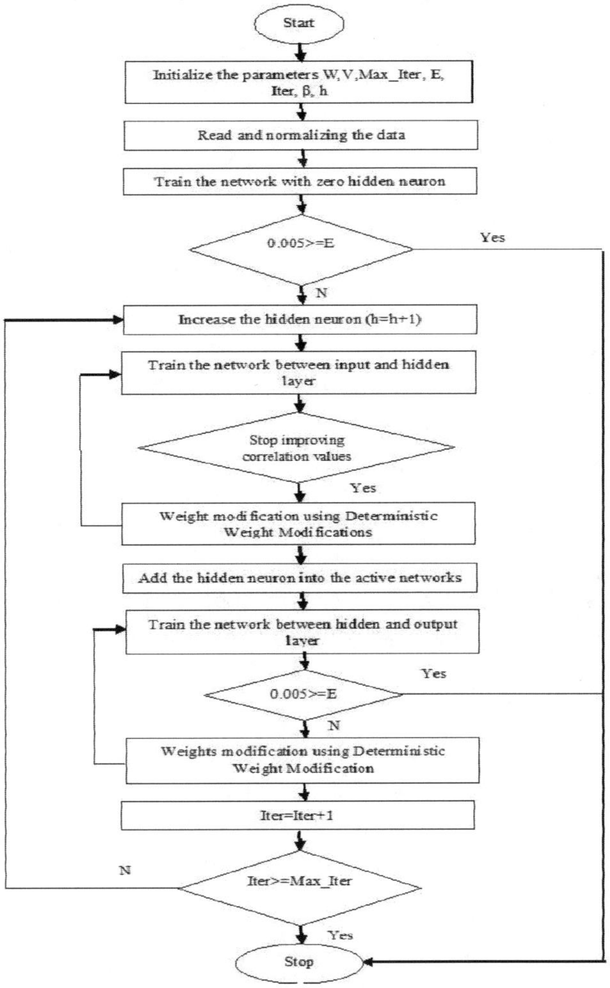

m . t_{po} is

Figure. 3.1 : Overview of the proposed DWM+CCNN prediction model

the target output and predictable output is $y_{po}(i)+\beta(t_{po}(i)-y_{po}(i))$. β must not be moreover near to 1. If not, it will cause unbalanced behavior. When β is set to 1, the predictable yield will become the objective value. The DWM has an improved the ability of global convergence rate and accelerate the convergence speed of the learning algorithm.

4.Experimental results and analysis

In this section, the efficiency of the proposed DWM-CCNN prediction is applied on two benchmark stock market indices datasets. Experimental outcomes are conducted with the help of MATLAB R2015b. The robustness of proposed DWM+CCNN prediction model is compared with conventional CCNN and BPNN.

4.1 Datasets collections

The data sets are covering the periods between January 3, 2005 and June 30, 2018. The proposed DWM-CCNN prediction model is applied to forecast the two Indian stock market indexes including Nifty 50 dataset and S&P BSE Sensex dataset. The stock market index prediction is a most challenging task in order to support investors to produce more profit. The dataset has many prices such as opening price, the highest price, the average price, low price and closing price.

Hence, four variables are considered as an input value to the prediction algorithm such as the Open Price (OP), High Price (HP), Average Price (AP), Low Price (LP) and then one variable is considered as a target output value such as closing price (CP). The data sets are collected from Bombay Stock Exchange (BSE) [22] and National Stock Exchange (NSE) [23] websites respectively. Each data set consists of 3350 trading days and it transformed into two phases, namely training phase and testing phase. From the datasets, 2850 data points are used for training part and the remains 500 data samples are used for the testing part.

4.2 Performance measures

The strength of the proposed DWM+CCN prediction method is calculated by statistical indicators such as Root MSE (RMSE), Mean Absolute Error (MAE), and Directional Symmetry (DS).

$$RMSE=\sqrt{\frac{1}{n}\Sigma_{i=1}^{N}(t_i-y_i)^2} \tag{21}$$

$$MAE=\frac{1}{n}\Sigma_{i=1}^{N}(t_i-y_i) \tag{22}$$

$$DS=\sqrt{\frac{100}{n}\Sigma_{i=1}^{N}d_i}, d_i=\begin{cases}1 & (y_i-y_{i-1})(t_i-t_{i-1})\\ 0 & otherwise\end{cases} \tag{23}$$

Where, t_i is represents the target value of given datasets, y_i is represents the predicted output of ANN prediction algorithm and N - represent the total number of data samples. The smallest value of RMSE and MAE are considered as the best performance of the time series analysis predictor. The DS present the exactness of the prediction model in term of percentage. In this study, the weights are initialized by small random values and termination conditions is set 0.005 (Error value). The maximum epoch is set as 2000 for all compared algorithms.

4.3 *Discussion*: The results of the proposed model that the DWM-CCNN is compared with existing standard CCNN and BPNN. In this research paper, single hidden layer is considered for stock market prediction and the quantity of hidden neurons is set 15 [24,25]. The experimental outcomes are obtained by the average values of 50 trials for comparisons in each algorithm.

Table 1 : Performance comparisons of the proposed prediction method

Methods	S&P BSE Sensex			Nifty 50		
	RMSE	MAE	DS(%)	RMSE	MAE	DS(%)
DWM-CCNN	**0.0185**	**0.0100**	**87.42**	**0.0096**	**0.0023**	**91.34**
CCNN	0.0626	0.0244	83.56	0.0124	0.0089	82.23
BPN	0.0975	0.0844	80.36	0.0139	0.0094	79.34

Figure 4.1 : Performance analysis of the prediction models for Nifty 50 datasets

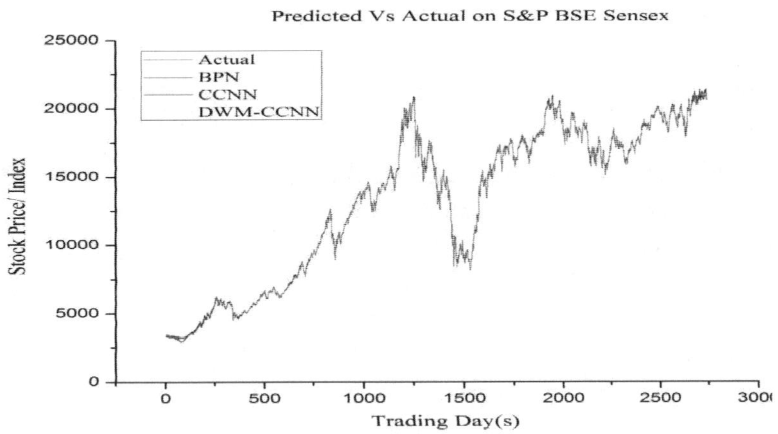

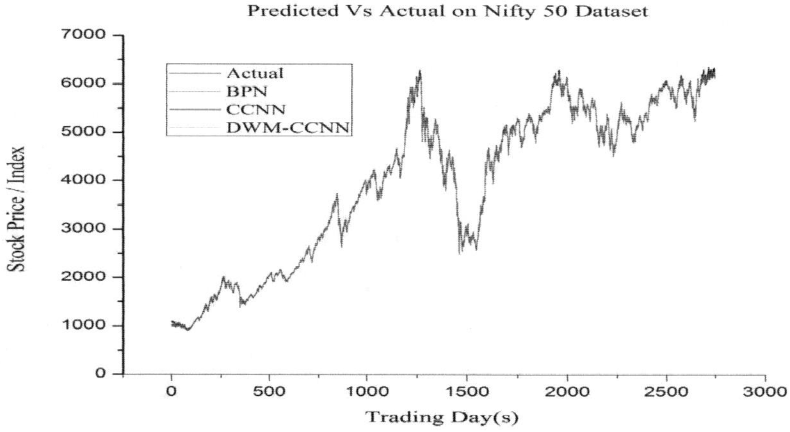

Figure 4.2 : Performance analysis of the proposed prediction models for S & P datasets

The statistical parameter values of the standard prediction algorithms are demonstrated in Table 1 and graphical representation shows in Figure 2 and Figure 3 for both S & P Sensex and Nifty 50 data sets respectively. The proposed model is produced higher prediction accuracy compared with CNNN and BPNN prediction methods. The optimal hidden neurons are an essential task which is determines the efficiency of the architecture. Hence, the proposed prediction model obtained the higher forecast accuracy and low error with 9 hidden neurons when applying for S & P Sensex and 8 hidden neurons when applying Nifty 50 datasets.
Commonly, the proposed system has reduced the slight enhanced prediction accuracy compared with CCNN and average enhancement accuracy compared with BPNN on both benchmark stock market indices datasets in terms of prediction accuracy and choosing appropriate hidden neurons. From the research works, we conclude that proposed method can prevent from the overtraining or undertraining of the learning process and produced higher prediction accuracy with less error.

5.Conclusions

The present research work is empirically applied to the stock market predication. The performance of DWM+CCNN is compared with standard CCNN and BPNN. The performance of proposed prediction model is analyzed using statistical measures such RMSE, MAE and DS. The proposed DWM-CCNN prediction algorithm is produced high convergence rate with optimal error, finding a suitable optimal number of hidden neurons and producing higher prediction accuracy. The proposed model has formed a significant enhancement with higher accuracy when compared with standard CCNN and BPNN. Future enhancement of this research work is applied to deep neural network (DNN) for stock market prediction in order to produce more efficient predication results. Because the conventional CCNN takes more computation time and produced low accuracy. Hence, the DNN is significant method that can be produced high accuracy when compared with normal neural network algorithms.

References

[1] Li F Qiao J Han H and Yang C 2016 A self-organizing cascade neural network with random weights for nonlinear system modeling *Applied soft computing* **42** 184-193
[2] Amalraj R and Dharmalingam M 2015 A work point count system coupled with back-propagation for solving double dummy bridge problem *Neurocomputing* **168** 160-178.
[3] Sheela K G and Deepa S N 2013 Review on methods to fix number of hidden neurons in neural networks *Mathematical Problems in Engineering*.
[4] Kwok T-Y and Yeung D-Y 1997 Constructive algorithms for structure learning in feedforward neural networks for regression problems *IEEE transactions on neural networks* **8** 630-645
[5] Madhiarasan M and Deepa S 2016 A novel criterion to select hidden neuron numbers in improved back propagation networks for wind speed forecasting *Applied intelligence* **44** 878-893
[6] Teoh E J Tan K C and Xiang C 2006 Estimating the number of hidden neurons in a feedforward network using the singular value decomposition *IEEE Transactions on Neural Networks* **17** 1623-1629
[7] Huang G Song S and Wu C 2012 Orthogonal least squares algorithm for training cascade neural networks *IEEE Transactions on Circuits and Systems I: Regular Papers* **59** 2629-2637
[8] Madhiarasan M and Deepa S 2017 Comparative analysis on hidden neurons estimation in multi layer perceptron neural networks for wind speed forecasting *Artificial Intelligence Review* **48** 449-471
[9] Kalaiselvi K Velusamy K and Gomathi C 2018 Financial prediction using back propagation neural networks with opposition based learning *Journal of Physics: Conference Series* **1142** 012008.
[10] Castellano G Fanelli A M and Pelillo M 1997 An iterative pruning algorithm for feedforward neural networks *IEEE transactions on Neural networks* **8** 519-531
[11] Xi L Muzhou H Lee M H Li J Wei D Hai H and Wu Y 2014 A new constructive neural network method for noise processing and its application on stock market prediction *Applied Soft Computing* **15** 57-66
[12] Vivarelli F Fariselli P and Casadio R 1997 The prediction of protein secondary structure with a cascade correlation learning architecture of neural networks *Neural Computing & Applications* **6** 57-62
[13] Ng S Cheung C and Leung S 2004 Deterministic weight modification algorithm for efficient learning IEEE International Joint Conference on Neural Networks (IEEE Cat. No. 04CH37541) IEEE
[14] Fahlman S E and Lebiere C 1990 The cascade-correlation learning architecture," in Advances in neural information processing systems
[15] Velusamy K and Amalraj R 2017 Performance of the cascade correlation neural network for predicting the stock price Second International Conference on Electrical, Computer and Communication Technologies (ICECCT), IEEE.
[16] Shultz T R and Fahlman S E Cascade-Correlation Encyclopedia of Machine Learning Editior, Springer US.
[17] Fahlman S E 1988 An empirical study of learning speed in back-propagation networks

[18] Treadgold N K and Gedeon T D 1997 A cascade network algorithm employing progressive RPROP International Work-Conference on Artificial Neural Networks Editior Springer

[19] Liang H and Dai G 1998 Improvement of cascade correlation learning *Information Sciences* **112** 1-6.

[20] Ng S-C Cheung C-C and Leung S-H 2004 Magnified gradient function with deterministic weight modification in adaptive learning *IEEE Transactions on Neural Networks* **15** 1411-1423

[21] Ben-Israel A and Greville T N 2003 Generalized inverses: theory and applications (Springer Science & Business Media).

[22] Wang J Wu L Choo K-K R and He D 2019 Blockchain Based Anonymous Authentication with Key Management for Smart Grid Edge Computing Infrastructure *IEEE Transactions on Industrial Informatics*

[23] Zhang Y Deng R H Liu X and Zheng D 2018 Blockchain based efficient and robust fair payment for outsourcing services in cloud computing *Information Sciences* **462** 262-277

[24] Sin-Chun N and Chi-Cheung S-H L 2004 Magnified gradient function with deterministic weight modification in adaptive learning *Neural Networks* **6** 1411-1423

[25] Dai W Wu J-Y and Lu C-J 2012 Combining nonlinear independent component analysis and neural network for the prediction of Asian stock market indexes *Expert systems with applications* **39**, 4444-4452.

UFMDRA: Uterine Fibroid Medicinal Drugs Review Analysis

Vineeta[1], Asha S Manek[2], Pranay Mishra[3]

[1]AMC Engineering College, Bengaluru, [2]RV Institute of Technology & Management, Bengaluru, [3]Hitachi Vantara, Bengaluru

E-Mail: vini.upmanyu@gmail.com

Abstract: Nowadays, enormous reviews are posted online by the consumers which provide related and required knowledge to the similar consumers. Such information is very much crucial for decision making and hence trendy among the web users. This information is very essential not only for potential consumers to make decisions but also for forecasting success and sustainability in commercial businesses. Online reviews on medicinal drugs are important for patients, medical representatives and medical industries. Reviewing medicinal drugs is challenging as sentiment analysis provides very little opportunity to discuss it. Collecting the reviews for uterine fibroid medicines from websites and analysing is a challenging process. An efficient Uterine Fibroid Medicinal Drugs Review Analysis (UFMDRA) model is developed with a decision tree algorithm which is trained and tested for different split ratios to obtain 100% accuracy. Experimental analysis results show that the proposed model has better classification performance in terms of accuracy compared to other classifiers.

Keywords: ***Decision Tree; Medicinal Drugs; Pre-processing; Review Analysis; Uterine Fibroid.***

1.Introduction

Most of the women suffer from problems of uterine fibroids which are non-cancerous tumors present around the uterus commonly found in the female pelvis. Presence of these fibroids ranges between 3.3 and 77% and varies with age. Fibroids are very common in childbearing women. The study conducted between July 2015 and August 2017, in a private hospital in the city suburbs, found that 11.6 % of the 4,487 women who visited the gynaecology outpatient department had uterine fibroids [1]. By the age of 50 most of the women have fibroids and underage of 20 [2] fibroids are rare. Studies suggest that this disease affected 60% of reproductive-aged women and during their lifetime 80% of women develop this disease [3, 4].

Fibroids can develop anywhere in the uterus and are classified according to their source of development. Myometrial/ Intramural Fibroids develop in and around the wall of the uterus.

Submucosal Fibroids grow under the surface of the uterine lining. Subserosal Fibroids grow exactly below the external layer of the uterus. Pedunculated Fibroids develop on a long stalk on the external and internal layer of the uterus.

The exact cause of fibroid development is still unidentified and believed to be caused due to a hormone named estrogen. Other reasons that are responsible for uterine fibroid can be age, obesity, eating habits and family history. Also, most of the time it is difficult to detect uterine fibroids through some symptoms unless physical examination is done. Fibroids get shrunk and show no symptoms for the women who are in menopause state. According to a recent study, fibroids shrink in women during premenopausal state also and an already existing fibroid doesn't increase the threat of developing a cancerous fibroid. Chances of getting other types of cancer in the uterus are also minimal if a woman already has fibroid. Fibroids can be cured by medicines but still in some cases where medication is ineffective, surgery is advised by the doctors. Various surgical procedures can be outlined as Hysterectomy, Myomectomy, Hysteroscopic resection of fibroids, Hysteroscopic morcellation of fibroids.

Doctors prescribe medicines such as Tranexamic Acid, Lupron, Ullipstical, Ulipristal Acetate, Anti–inflammatory medicines such as ibufren, contraceptive pills and gonadotropin for the symptoms and/or shrink of fibroids. People bearing uterine fibroids who consume these medicines may have various side effects. Depending upon their experiences people write their feedback/reviews online on these medicinal drugs. These reviews can be used by other patients who have been prescribed the same medicines to make up their minds about it. Also, these reviews are helpful to pharmaceutical companies and doctors who prescribed those drugs. Reviews are given in terms of comments where the users write their experiences after taking the drug and give some points as rating. But it is observed that sometimes ratings given by users contrast with the comments given by the same users. Therefore, Sentiment Analysis is necessary to find the exact meaning of comments where the comments can be categorized as positive, negative or neutral depending on its sentiment.

Two drugs Tranexamic acid and Lupron have been used in this paper for analysis of user reviews. Different machine learning models are used to classify which drug is best from the users' perspective. Drug reviews are scraped and collected in comma separated value (csv) format along with ratings provided by the users. Comments of medicine consumers are classified as positive, negative or neutral.

This paper is structured in the following way: Section 2 describes related work, section 3 explains the motivation behind the presented work, the system design and implementation is presented in section 4, obtained results are shown in section 5 and section 6 summarizes the paper.

2.Related Work

Review analysis in recent years has not provided much information about uterine fibroid medicinal drugs so it is a challenge to analyse its reviews. Recent investigations on medicinal drug reviews recommend that online reviews and blogs are vital sources of information for sufferers, distributors and healthcare professionals. The main challenge is extracting the related particulars about medicinal drugs from the online sources and investigating them. An ensemble of W-Bayesian Logistic Regression and Support Vector Machine (W-LRSVM) model is proposed by Asha S Manek et. al., to perform review analysis on medicinal drugs [5]. The W-LRSVM model collects the information from the web by using a web crawler. The model was trained by collecting reviews from drug.com for 5 drugs - Escitalopram, Citalopram, Lisinopril, Oxycodone and Lyrica. Lisinopril is used to regulate blood pressure, Citalopram and Escitalopram is used for depression; Oxycodone drug is used to treat severe pain whereas Lyrica drug is for neuropathic pain. The W-LRSVM model outperformed with 97.46% accuracy after training it for different split ratios.

Nowadays, consumers are benefited with online reviews posted on the Web which provides required information highly essential for decision making. This valuable information is important for eventual customers as well as for anticipating benefits and improvement in business. However, in spite of availability of various algorithms, there is no particular method that can provide a solution for analysis of the sentiments. In paper [6], authors proposed a Support Vector Machine (SVM) classification approach for analysis of large movie review data set. The results showed classification performance in terms of reduced error rate and accuracy using Gini Index based feature selection method.

Xuan, J., et al., in paper [7] analysed 328 medical literature of uterine fibroid treatment carried out on western medicines that primarily included leuprolide (GnRH-a), danazol (androgen), gestrinone (progesterone), mifepristone (progesterone receptor antagonist) and other medicines. The collected literature data of uterine fibroid were reviewed and established that the western medicines were used in circumstances like (i) patients having uterine fibroids without symptoms; (ii) patients having normal menstrual cycle of 22–34 days; (iii) patients aged between 25 and 49 years; (iv) patients having uterine of size less than ten weeks of pregnancy, tumour diameter between 2 cm and 5 cm and patients without menopause were also considered. This paper literature observed that among the 328 selected medical literature reports, there were 52 cases (15.9%) of leuprorelin, 60 cases (18.4%) of danazol, 44 cases (13.4%) of gestrinone, and 78 cases (23.8%) of mifepristone. Authors have used data mining technology to examine the prescriptions for uterine fibroids treatment in recent 15 years. The different combinations of drugs, frequency of drug and other prescriptions for the treatment of uterine fibroids were also examined.

Han, Y., Liu and M. and Jing, W., performed aspect-level drug review sentiment analysis by proposing a data set SentiDrugs that contains one or more targets. A proposed Pretraining and Multitask Learning model based on Double BiGRU (PM-DBiGRU) in which the pretrained weight is used to initialize related weight. The target-specific aspect-level drug review is represented by applying two BiGRU networks. Further, the domain knowledge is transferred from the short text-level drug review data set. The PM-DBiGRU model verifies that bidirectional semantic information of target plays an important role for classification of sentiment polarity in drug reviews and domain knowledge improves the aspect-level classification performance [8].

Vijayaraghavan, S. and Basu, D., analyzed reviews of various drugs which have been reviewed in form of texts and have also been given a rating on a scale from 1-10. The proposed system obtained the data set from a repository of UCI machine learning which had 2 data sets: train and test (split as 75-25%) to investigate the reviews for different conditions and use different words to impact the ratings of the drugs. The authors implemented supervised machine learning classification algorithms for predicting the class of the rating using the textual review by embeddings Term Frequency Inverse Document Frequency (TFIDF) and the Count Vectors (CV). The proposed model performed training on the most popular conditions such as "Birth Control", "Depression" and "Pain" within the data set [9].

3.Motivation

Review analysis identifies many applications and it is very beneficial for reading consumer mindset/ viewpoint about products/services. In the related work section, literature survey has done on quite a few issues of review analysis of medicinal drugs. In contrast, still there is a scope to enhance the opinion detection and classification performance for medicinal drugs. Most of the analysts have experimented on common domains for review analysis such as electronic gadgets, movie shows and restaurants reviews but not much on health and medicinal domains specifically on Uterine fibroid medicines. We focused to overcome this problem by developing an effective method for review analysis using uterine fibroid medicine review data sets. Several drugs are available nowadays for uterine fibroids treatment. But most of the drugs have side effects such as affecting ovary ovulation, intervention with the role of progesterone and estrogen receptor, growth of vascular endothelial in fibroid tissue and size of fibroid enlargement. The vital challenge to analyse reviews of uterine fibroid medicines by scrapping the reviews of consumers and applying proper pre-processing filters, effective code and data transformation methods. The performance of this work shows that analysis of uterine fibroid medicines reviews efficiently improved by training the proposed model with Decision Tree classifier.

4.System Design & Implementation

The proposed system for uterine fibroid medicines reviews is carried out in four steps as shown below in Figure 4.1.

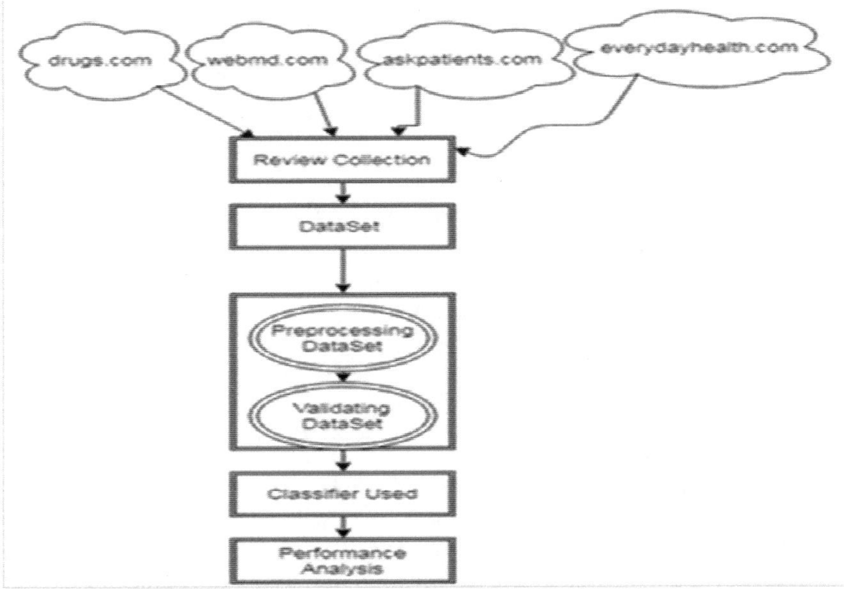

Fig 4.1: Proposed system for UFMDRA Model

4.1 Review Data Collection

The first step in the proposed system is data collection. Data collection is done from the different websites namely drugs.com, everydayhealth.com, askpatients.com and webmd.com. Scraping is used for collecting reviews and ratings from the above websites resulting in our own data set. Python BeautifulSoup is used for web scraping and a total 571 reviews have been scrapped. After scraping, the reviews are collected in a comma separated value (.csv) file to make a data set with drug name, comment given by user and ratings given by user.

4.2.Data Preprocessing

As the data set contains positive, negative and neutral reviews in the comment section and ratings given by users are also present, preprocessing is done. Preprocessing step consists of chopping up character sequence into tokens (tokenization), changing into lower cases, lemmatization, refining by removing stop words and collection of sentiment words. In the rating column missing values are taken care of by replacing with zero (0) values as such cases are very less. Valence Aware Dictionary for Sentiment Reasoning (VADER) Sentiment Analysis technique based on lexicons of sentiment words is used for the proposed system [10]. Each of the words in the lexicon is rated in terms of polarity scores that how much it is positive, negative or neutral. VADER is valence based where intensity of the sentiment is taken into account. Same sentences written in different ways might have different polarity scores that are very well taken care of by VADER. The attributes are extracted and then preprocessed to generate the final data set. The selected attributes for the proposed model is as shown in Table I.

TABLE 1: SELECTED ATTRIBUTES FROM DATA SET FOR THE PROPOSED MODEL

drugname	ratingSentimentLabel	cleanReview	vaderSentiment	ratingSentiment
review	vaderReviewScore	ratings	vaderSentimentLabel	

4.3. *Data Validation*

The validation of preprocessed data generally used for model construction is addressed below. The Split Validation is carried out to assess the performance of a learning operator and the data set is split into training and test. The Split Validation operator can use several types of sampling methods for building the subsets. Cross validation is used to build the proposed model which avoids the sampling bias. The various cross validation methods are k-Fold, Leave One Out Cross Validation (LOOCV), Leave One Group Out Cross Validation (LOGOCV), Nested Cross Validation, Time Series Cross Validation and many more. The proposed system is built by using k-Fold cross validation technique.

4.4. *Modeling*

The approach of using Machine Learning algorithms is based on the desired result of the algorithm or the type of input available during training of the model. The Machine Learning classifiers used in proposed work are Random Forest (RF), Decision Tree (DT), K- Nearest Neighbor (KNN), Logistic Regression (LR), Naïve Bayes (NB), Support Vector Machine (SVM) and Stochastic Gradient (SGD). The ensemble of Logistic Regression, Decision Tree and Support Vector Machine is also used in different ways to analyze the performance of the proposed model. Classifiers are trained on different train/test split ratios with and without tuning of hyper parameters. For k-fold cross validation, different numbers of splits are taken and analyzed to determine the performance of the model.

5.Results & Performance Analysis

Uterine Fibroid Medicinal Drugs Review Analysis (UFMDRA) model is designed using python programming and performance is evaluated by applying various classifier algorithms. Evaluation related terminology used is accuracy which is calculated by varying the split ratio from 0.7 to 0.9 and k-Fold cross validation. In the review analysis of the UFMDRA model, the terms True Negative (TN), True Positive (TP), False Negative (FN), False Positive(FP) compare results of the algorithms used. Positive and Negative refers to the algorithms prediction while True and False indicates the prediction made respective to the observation.

First the reviews are collected from drugs.com, everydayhealth.com, askpatients.com and webmd.com using the scrapping program for 2 uterine fibroid related drugs- Tranexamic acid and Lupron. The next step is to analyse the collected and preprocessed reviews. The reviews of uterine fibroid drugs given as input to the UFMDRA model and performance is compared with different combinations of classifiers depicted in Table II, III and IV.

Test Size	Classifiers Used with Accuracy in %						
	Random Forest (RF)	*Decision Tree (DT)*	*k-Nearest Neighbor (KNN)*	*Logistic Regression (LR)*	*Naive Bayes (NB)*	**Support Vector Machine (SVM)**	**Stochastic Gradient (SGD)**
0.30	98.25	100	94.76	95.34	98.83	94.76	98.25
0.20	100	100	100	98.26	100	98.26	98.26
0.10	98.27	100	98.27	93.10	96.55	91.37	94.82

TABLE 2: PERFORMANCE OF DIFFERENT CLASSIFIERS ON DEFAULT PARAMETERS

The data set is divided into training and testing data sets first. The proposed system is trained using the training set and the performance is evaluated in terms of accuracy for three test sizes 10%, 20% and 30% using default parameters and after tuning hyperparameters of algorithms as illustrated in Table II and Table III respectively.

TABLE 3: PERFORMANCE OF DIFFERENT CLASSIFIERS AFTER TUNING HYPER PARAMETERS

Test Size	Classifiers Used with Accuracy in %						
	Random Forest (RF)	*Decision Tree (DT)*	*k-Nearest Neighbor (KNN)*	*Logistic Regression (LR)*	*Naive Bayes (NB)*	**Support Vector Machine (SVM)**	**Stochastic Gradient (SGD)**
0.30	99.41	100	96.51	99.41	-	94.76	100
0.20	100	100	100	100	-	100	100
0.10	98.27	100	98.27	98.27	-	98.27	96.55

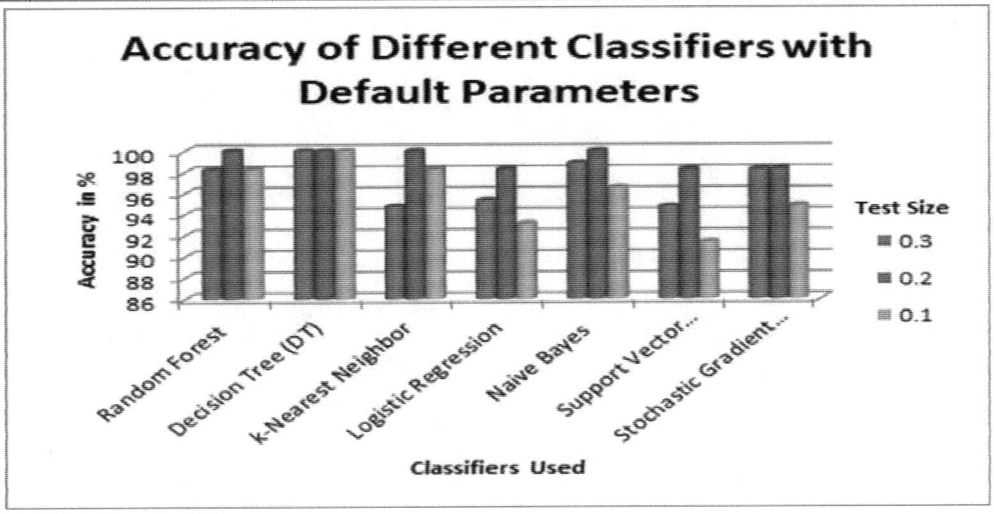

Fig 5.1: Comparison of all Classifiers of various test size with Accuracy with Default Parameters

The graphical analysis of performance of all classifiers shown in fig 5.1 shows that the Decision Tree algorithm outperforms with 100% accuracy in all cases. All the classifiers have shown best results with test size of 20% with not much changes in accuracy for Random Forest Algorithm. There is an increment in accuracy if the test size increased from 10% to 20% for all classifiers but there is an decrement in accuracy if the test size increased from 20% to 30% for all classifiers except Decision Tree algorithm.

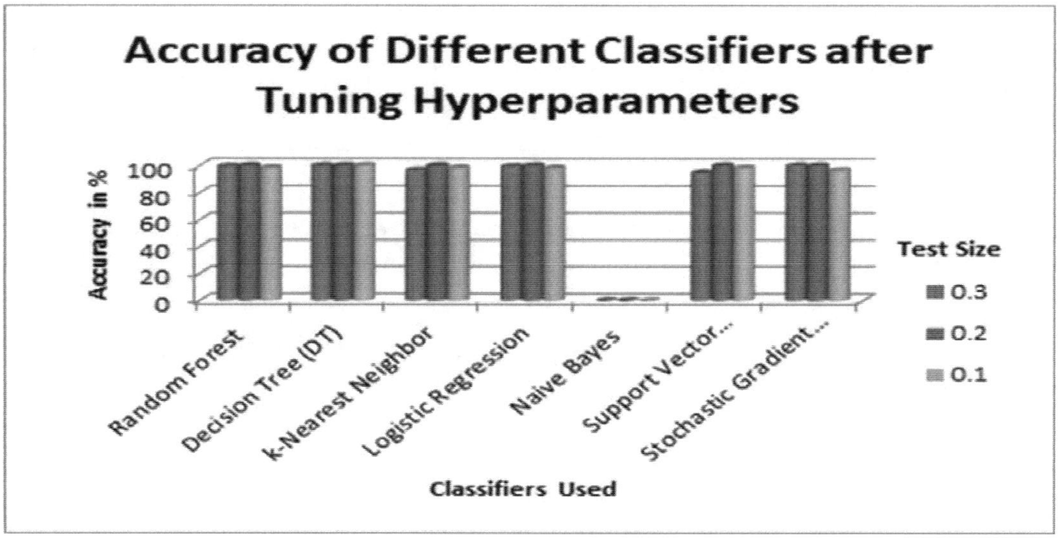

Fig 5.2: Comparison of all Classifiers of various test size with Accuracy after Tuning Hyperparameters

The graphical analysis of performance of all classifiers after tuning parameters shown in Fig 5.2 shows that the Decision Tree algorithm outperforms with 100% accuracy in all cases irrespective of test sizes. In this case also, there is an increment in accuracy if the test size increased from 10% to 20% for all classifiers but there is not much decrement in accuracy except Support Vector Machine algorithm if the test size increased from 20% to 30%. The graphical analysis shows that Decision Tree out performs with other compared classifiers on the given data set irrespective of test sizes or refinement of hyperparameters.

The reviews of uterine fibroid medicines given to the UFMRA model and performance is compared with different combinations of classifiers like Logistic Regression + KNN , Logistic Regression + SVM , KNN + SVM and Logistic Regression + KNN + SVM . The graphical analysis of performance (i.e Accuracy) is shown in Fig 4. With the ensemble of classifiers, we obtained 96.33% 94.77%, 96.33% and 94.95% for Logistic Regression + KNN, Logistic Regression + SVM, k-N + SVM and Logistic Regression + KNN + SVM classifiers respectively with k-Fold splits equals to 20 as shown in Table IV.

TABLE 4: PERFORMANCE OF ENSEMBLED ALGORITHMS

k-Fold Splits	Classifiers Used with Accuracy in %			
	Logistic Regression + K- Nearest Neighbor (LR+KNN)	*Logistic Regression + Support Vector Machine (LR+SVM)*	*K- Nearest Neighbor + Support Vector Machine (KNN+SVM)*	*Logistic Regression + K- Nearest Neighbor + Support Vector Machine (LR+KNN+SVM)*
10	95.97	94.74	95.97	94.92
15	96.14	94.75	96.14	94.75
20	96.33	94.77	96.33	94.95

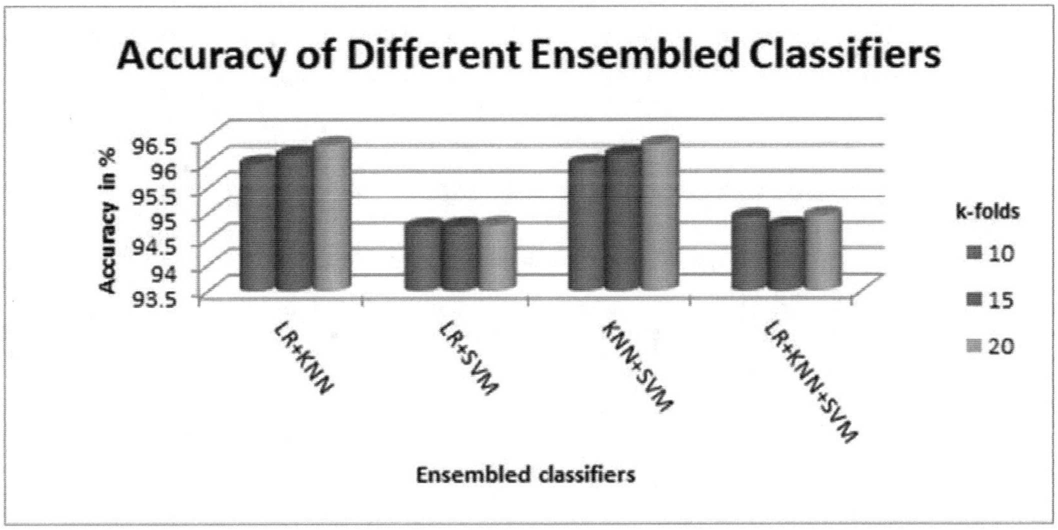

Fig 5.3: Performance Analysis in terms of Accuracy of different Ensemble of Classifiers

5.1.Comparison of UFMDRA system with work [11]:

In paper [11], a combination of context-based sentiment analysis using Ngram and tf-idf word vectorization method approach is presented using Naïve Bayes and Random Classifiers to predict user review emotion with accuracy of 89%. With the Decision Tree classifier, UFMDRA outperformed with 100% accuracy as shown in Table V.

TABLE 5 : COMPARISON OF UFMDRA WITH REFERENCE[11]

Model	UFMDRA	Comparison with Work [11]
Data set	Scraping data from online websites drugs.com, everydayhealth.com, askpatients.com and webmd.com	UCI ML Drug Review & Online Pharmaceutical Review sites
Best Classifier	Decision Tree	Naive Bayes and Random Classifier
Accuracy	100%	89%

6.Conclusion

In this work, we generated a data set from drugs.com, everydayhealth.com, askpatients.com and webmd.com using the scrapping program for 2 uterine fibroid related drugs- Tranexamic acid and Lupron. We explored seven supervised learning models: Random Forest , Decision Tree-Nearest Neighbour , "Logistic Regression , "Naive Bayes , Support Vector Machine and Stochastic Gradient for the purpose of determining the sentiment polarity of uterine fibroid drug reviews. We found that the Decision Tree model performed with the highest accuracy of 100%. In future a work user interface can be designed so that any consumer can easily find the reviews class. The UFMDRA model can be applied to analyse any drug reviews with adding and scrapping more data of uterine fibroid related medicines.

References

[1] http:// www.newindianexpress.com/cities/ hyderabad/2017 /nov/12 / uterine-fibroids widely-prevalent-in-city-study- 698910.html

[2] https:// www.indiatoday.in/lifestyle / health/ story/story/uterine- fibroids/ 1/ 390873-300316-2014 -09-24]

[3] Myers SL, Baird DD, Olshan AF, Herring AH, Schroeder JC, Nylander-French LA, et al. Self-report versus ultrasound measurement of uterine fibroid status. J Womens Health (Larchmt) 2012;21:285–293. [PMC free article] [PubMed] [Google Scholar]

[4] Ryan GL, Syrop CH, Van Voorhis BJ. Role, epidemiology, and natural history of benign uterine mass lesions. Clin Obstet Gynecol. 2005;48:312–324. [PubMed] [Google Scholar]

[5] Manek, A.S., Pandey, K., Shenoy, P.D., Mohan, M.C. and Venugopal, K.R., 2015, December. Classification of drugs reviews using W-LRSVM model. In 2015 Annual IEEE India Conference (INDICON) (pp. 1-6). IEEE.

[6] Manek, A.S., Shenoy, P.D., Mohan, M.C. and Venugopal, K.R., 2017. Aspect term extraction for sentiment analysis in large movie reviews using Gini Index feature selection method and SVM classifier. World wide web, 20(2), pp.135-154.

[7] Xuan, J., Deng, G., Liu, R., Chen, X. and Zheng, Y., 2020. Analysis of medication data of women with uterine fibroids based on data mining technology. Journal of infection and public health, 13(10), pp.1513-1516.

[8] Han, Y., Liu, M. and Jing, W., 2020. Aspect-Level Drug Reviews Sentiment Analysis Based on Double BiGRU and Knowledge Transfer. IEEE Access, 8, pp.21314-21325.

[9] Vijayaraghavan, S. and Basu, D., 2020. Sentiment Analysis in Drug Reviews using Supervised Machine Learning Algorithms. arXiv preprint arXiv:2003.11643.

[10] https://www.nltk.org/_modules/nltk/sentiment/vader.html

[11] Tumu, P., Manchenasetty, V. and Rege, M., 2020. Context based sentiment analysis approach using n-gram and word vectorization methods, Issues in Information Systems, 21(3), pp.59-65.

Performance of particle swarm optimization bin packing algorithm for dynamic virtual machine placement for the consolidation of cloud server

[1] C. Pandiselvi [2] Dr. S. Sivakumar

[1] Research Scholar and Assistant Professor of Computer Science, CPA College, Bodinayakanur

[2] Principal, CPA College, Bodinayakanur

Abstract. Infrastructure as a service offered by the cloud computing is one of the most important service. It allows physical machines to get virtualized by creating many instances of virtual machines. Mapping virtual machines on physical machine has become the major challenge in cloud data centres. The dynamic virtual machine placement methods are used to solve this issue with objectives like maximizing the resource utilization, minimizing the energy consumption and maximizing the scalability of data centres. In this paper a virtual machine placement-based bin packaging algorithm is proposed and analysed with four different fitness strategies to obtain the optimal solution. The unimodal (Sphere, Step) and multimodal (Graywang and Rastridge) benchmark functions are used with proposed algorithm for the analysis and obtain the quantitative measurements. The results show that optimizing the mass of particles using the best fitting strategy reduces the energy consumption, resource utilization and improved the scalability of data centres.

Keywords: *Virtual machine placement, Energy consumption, Resource utilization*

1. Introduction

Cloud computing based on internet has the powerful architecture that the user has to pay only for the services they have used. The three basic components in cloud computing are (i)Client computer, where the end user can interact with cloud, (ii) Distributed servers, where the servers are distributed among different places, (iii) Data centres, where they are the compilation of servers [1]. In cloud datacentres the cloud service providers have been attracted by millions of clients, who are sharing resources using servers. By adding virtualization technologies to the servers, it enables virtual machines (VM) live migration and allows the VM to be freely moved among physical machines (PM). Therefore, several VM can be packed on a single PM known as server consolidation [2]. Server Consolidation allows datacentres to improve resource utilization and energy efficiency by placing the virtual machines on the required capacity of the PM [3].

In placing the VM on PM leads to inefficient utilization of the resources, where VMs are periodically reallocated using live migration according to their current resource demand. In order to minimize the number of PM, workload should be balanced and the idle servers are switched to low-power modes which reduce the overall energy consumption and cost. There are many studies has been committed to solve the VM placement, load balancing, energy consumption and cost profit in cloud [4]. We contribute three steps in this research work, (i) The entire service request are considered as particles, grouped ordered to form a group and hunt for good solution. A virtual machine placement technique-based bin packing algorithm (VMBP) to get an optimal solution, (ii) it is placed in bins, where bins represent a PM and items represents the VM to find the efficient placement of VM. The process of finding solution using particle swarm optimization technique (PSO) embed bin packing (BP) with best fit strategy leads to resource utilization and energy consumption. (iii) The benchmark functions such as Sphere, Step, Graywang and Rastridge are applied to evaluate the VMBP algorithm. It uses the exponential decreasing inertia weight, so that to replace it to the linear inertia weight for autonomous groups particles and to get better balance global search and local search of the algorithm for improving the algorithm's ability [11].

The remainder of the paper is organized as follows, in section 2, related works are discussed, in section 3, the system model for PSO and BP algorithm is explained, and section 4 presents the details of proposed algorithm VMBP followed by section 5 presents the simulation setup and results obtained. In section 6 concludes the paper with discussing the results and future research directions.

2. Related Works

An energy organization system is planned for the housing zone based on the experimental algorithm to attain the load gain in grid area network and a hybrid genetic-wind-driven (GWD) algorithm is proposed for the load of single homes and multiple homes. There are two types of instance is taken for consideration that single is on-demand case and added is reserved case, the proposed pricing schema algorithm, which is used to reduce the cost of cloud data centres [5]. In the Integration and green camp tutorial, a spatial awareness reduction system (SAPFT) for VM migration was proposed [6]. Energy optimization is performed using an numeral quadratic polynomial (IQP) with linear and quadratic constraints for the data center [7]. The information focus on management and configuration process is designed to be an optimization problem. Because the problem is unpredictable, a cross non-natural bee and group optimization algorithm is projected to optimize the cloud using smart tool resources [8][9], a VM placement optimization algorithm for location identification, to reduce the number of PM identifications. Effective and efficient resource management in grid computing [9].

A data center provides a virtualized stage to switch the vast volume of facility needs from the user. They need properties to run their requests with virtual machines, which can be provisioned and achieved by on-demand. The facts focus administrator must reply to numerous on-demand resource needs by defining where VM are located and how the possessions are owed to them. Assignment of VM is the most exciting part in fog datacentre [10].

3. System Model

The virtual machine placement problem in datacentre is formulated as bin packing problem of placing the VMs. Many research works focus only on single criterion for virtual machine placement, but real-world problems require multi objective situation, therefore considered the problem of VMP is formulated as multi objective optimization problem and simultaneously optimize resource wastage and energy consumption.

3.1. Particle Swarm Optimization problem (PSO)

The PSO method is a swarm-intelligence-based experimental technique for the answer of optimization tricky. It is unpretentious, optimistic and controlling technique. It duplicates the common performance of fishes and birds [12]. In PSO each individual possible solution can be modulated as a particle that moves through the problem in search space S^n [13,14]. The position of each particle is determined by the vector $Y_i \epsilon S^n$ and its movement by the velocity of the particle $V_i \epsilon S^n$, as shown below,

$$Y_i^{(z+1)} = Y_i^{(Z)} + V_i^{(Z+1)} \quad \text{.......(1)}$$

Let $Y_i^{(Z+1)}$ denotes the position of ith particle at iteration (z+1). $V_i^{(Z+1)}$ is the velocity of ith particle at iteration (z+1). The position of the particle is changed by adding a velocity $V_i(z)$ to the current position. For each particle, fitness function is based on its own experience (the decisions that it has made so far and the success of each decision) and the knowledge of the performance of other individual particle in its neighborhood. Since the relative importance of these two factors can vary from one decision to another, it is reason to apply random weights to each part, and therefore the velocity will be determined in below equation *(2)*.

$$V_i^{(z+1)} = w * V_i^{(z)} + c1 * rand(r1) * (pb - Y_i^{(z)}) + c2 * rand(r2) * (gb - Y_i^{(z)}) \quad \text{..........(2)}$$

where *c1, c2* are two positive numbers and *rand (r1), rand (r2)* are two random numbers.

The velocity update equation in *(2)* has three major components,

- The first component *w* referred to as "inertia," It refers to the tendency of the particle to continue in the same direction it has been traveling.
- The second component is the best position of the particle pb: (whose corresponding fitness value is called the particle's best), scaled by a random weight *c1 * rand (r1)*.
- The third component is the velocity update equation of the best particle position found by any swarm gb: (whose corresponding fitness value is called global best), scaled by another random weight *c2 * rand (r2)*.

3.2. Bin Packing problem (BP)

The tricky can be seen as bin padding problem with adjustable bin sizes, where boxes represent the PM and objects are the VM to be packed. Bin padding tricky is used to parcel the extreme number of VM into a minor number of PM [15,16]. The main purpose of bin padding is to avoid consumption of resources, save energies and reduce price. BP problem can occur in two dimensions therefore in that there are two solutions, (i) Exact called heuristic solutions, are very complex and expensive in terms of computing time, (ii) Approximation called meta heuristic solutions are reliable and the results are obtained very fast.

The algorithms used to get this meta heuristic solutions are first fit, next fit, best fit and worst fit [17]. The BP algorithm with best fit strategy has the items are placed in bin with the order they arrived. Therefore, place the next item in bin where its capacity is closest. Next check if the item does not fit in any bin then open a new bin and place the item in that bin, hence the BP algorithm is used to pack the maximum number of VM into a small number of PM [18]. The main purpose of bin packing is to avoid wastage of resources and save energies.

The item and bins are used to minimize the number of bins, where item $P= \{p_1,p_2,p_3....p_n\}$ and bins $B = \{b_1,b_2....b_m\}$. Formally a charting is possible for all element have been successfully mapped to the bins, for all i, $g(i) \neq Ø$ and united request of resources is within the boxes capacities: for all k,j $g(k,j) \leq C_{j,k}$.
A solution of 2D bin packing problem can be represented by $F = \{f_1,f_2...f_i\}$, where F can be represented as $F_i = \{ (p_1,b_1), (p_2,b_2).....(p_n,b_m)\}$. The concluding optimization function can be Minimize $\sum_{j=1} Y_j$

where Y_j represents objective function minimizes the total number of bins used.

$$\sum_{i=1} W_{i,j} R_{j,k} \leq C_{j,k}$$

weight of the item $W_{i,j}$ placed in bin $R_{j,k}$ should not exceed bin size capacity $C_{j,k}$.

$$X_{i,j} = \{1, \text{ if } VM_i \text{ was placed on } PM_j$$

$$0, \text{ otherwise}$$

VM_i are the items was placed in a PM_j of bin.

$$Y_j = \{1, \text{ if} \sum_{i=1} x_{i,j} > 0$$

$$0, \text{ if} \sum_{i=1} x_{i,j} = 0$$

checks whether PMj is lively, if there exists more than one VM then yi = 1.

$$\forall i \in \{1,2,...n\}, \forall j \in \{1,2,...m\}, \forall k \in \{1,2,...(m,n)\}$$

demonstrates particular meaning for area of the variable i,j,k.

4. Virtual machine placement technique-based bin packing algorithm (VMBP)

In cloud data center, there are huge amount of service requests from the user, the cloud service provider who respond to the service request want resources to path their requests by private virtual machines, which can be provisioned and accomplished by on-demand.

However, the usage of resources and consumption of energy is increased so, to reduce the usage of resources the entire service request from the cloud user is considered as particles. In PSO technique, the number of units mix to make a group. The units initialized by assigning the random position of the unit. In every iteration each unit is updated by two" *best*" values, the one is the best position of the particle that is the fitness value of particle's best called *pbest*, as well as the another is the best position of the swarm that is the fitness value of particle's best called g*best* is found. Then compare the fitness value of pbest value better than *gbest* if it is true, set the *pbest* value as *gbest.*

Then update speed and location of the unit, repeat the iteration until the optimal solution is obtained. The updated unit position referred as items and placed in bin. Then, if the item sizes are greater than bin sizes then the VM are placed in bins, an efficient virtual machine placement is done are shown in Fig.4.1.the structure diagram of VMBP algorithm.

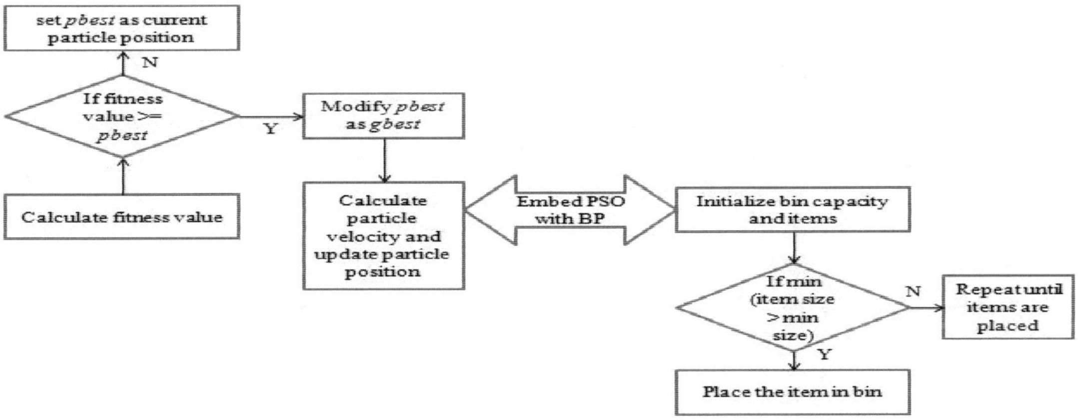

Figure 4.1: Structure diagram of VMBP algorithm

Step 1: For each particle

Initialize Y_i *and* V_i

Calculate *aptness value*

If *aptness value* better than *pbesti*

Set *pbesti* as *current particle position*

End

For each particle

Choose the *pbesti value* with *best fitness value* and set as *gbesti*

Calculate *particle velocity*

Then Update the particle position

Figure 4.2: VMBP algorithm pseudo code

The Pseudo code for VMBP has two steps is shown in Fig.4.2,

Step 1: Each unit has approximately place and it shows a probable explanation of the aptness function in search space where search space is a set of all probable positions. Each unit i maintains two routes in search space where, commercial route $Y_i = [y_i1, y_i2...y_id]$. Speed route describes the measure of a unit where, speed route $V_i = [v_i1, v_i2...v_id]$. *pbest* of unit i in d-dimension search space is represented as $pbesti = [p_i1, p_i2, p_i3, ..., p_id]$ *and gbest* in d-dimension search planetary is represented as $gbesti = [g_i1, g_i2, g_i3, ..., g_id]$.

Step 2: Each particle is updated as items in bin packing algorithm with best fit statergy. The best fit allocates the items to a bin which has the smallest sufficient space among the free available space. The minimum item is calculated using the formula min(item size > bin size) if found then place it to that bin or leave the items and check for further items. Repeat the process until the items are placed. The process of finding possible solution using PSO algorithm and embedding bin packing with best fit strategy leads to resource utilization and energy consumption.

5. Simulation Setup

The VMBP algorithm is used to solve VMP problem. In PSO technique a number of particles integrate to make a swarm and search for good solution. The efficiency in finding solution for benchmark functions with a lower number of particles and iterations leads to extremely simple procedure, easy implementation and very fast convergence rate. The parameters used in this algorithm are the '*number of particles*', where PSO take real numbers as particles, it can be set as (x1, x2, x3...X10), '*dimension of particles*' which is taken for 30 dimensions for each benchmark functions, '*range of particles*' determined as per the benchmark functions and '*Vmax*' determines the maximum velocity change of one particle during one iteration. The updated particle velocity of each particles is considered as the PM used in cloud. Therefore, by using the bin packing algorithm with best fit strategy the PM is placed in VM, here occurs resource utilization. In order to test the VMBP algorithm, the number of service request from the cloud user is difficult to handle, where benchmark functions are used to evaluate the algorithm in an efficient manner. Four well known unimodal and multimodal functions parameters are listed in the Tab.1.

Function Name	Dimension	Function Type	Search Range
Sphere	30	Unimodal	[-5.12 ,5.12]
Step	30	Unimodal	[-100,100]
Graywang	30	Multimodal	[-600,600]
Rastridge	30	Multimodal	[-15,15]

Table 1: Parameters of unimodal and multimodal functions

These functions are used to test good cases because of their nonlinearity and oscillation around the optimal solutions, so there exists a high probability for each optimization technique to trap into local optima. There are two types of benchmark functions namely unimodal functions and multimodal functions. The function having a single optimal solution is unimodal i.e., Sphere and Step. Sphere function contains no local optima and provides a smooth gradient towards global optima and Step function contains infinite number of global minima. Multimodal functions have two are more optimal solutions i.e., Graywang and Rastridge. Graywang function has many widespread local minima, which is regularly distributed and Rastridge function has several local minima and the locations of minima are regularly distributed [19, 20].

5.1 Experimental Result

To evaluate the multi objective algorithm VMBP with objectives like the resource utilization, energy consumption and scalability of data centres the open-source integrated development environment NetBeans IDE is used. The first part of this section discusses about the resource utilization by decreasing the number of active PM. The next part of the section discusses about the scalability of datacentres. In the rest of the section discuss about the consumption of energy using benchmark functions fitness value.

Resource utilization: In cloud datacentres each user requests single or additional applications with an predictable value of deal that needs a certain number of resources. The datacentre respond to the user desires for VMs are to be located to existing resources. The available resources are to be placed as well as resource utilization also needed for an efficient placement of VM. Therefore, in VMBP algorithm the number of particles taken to make a swarm is updated with position and velocity and the updated particles are taken as items for which they are placed in bin using best fit strategy. The minimum number of ten particles is taken for experiment is considered as ten cloud user requests. Using the algorithm VMBP the ten-user request for PM strives to consolidate ten PM in nine VM resulting in higher resource utilization is shown in Tab.2.

User Request (UR)	*Number of PM*	*Resources allocation in VM*
UR1	PM1	VM2
UR2	PM2	VM3
UR3	PM3	VM5
UR4	PM4	VM4
UR5	PM5	VM0
UR6	PM6	VM1
UR7	PM7	VM7
UR8	PM8	VM9
UR9	PM9	VM8
UR10	PM10	VM6

Table 2: Resources allocation for different user request in cloud

The number of physical machines is taken as X axis and the number of virtual machines is taken as Y axis. In VMBP algorithm the minimum bin is calculated and fits the item to the bin shows the placement of virtual machines is done. Therefore, result proves that effectively resource utilization takes place in VMBP placement algorithm are shown in Fig.5.1.

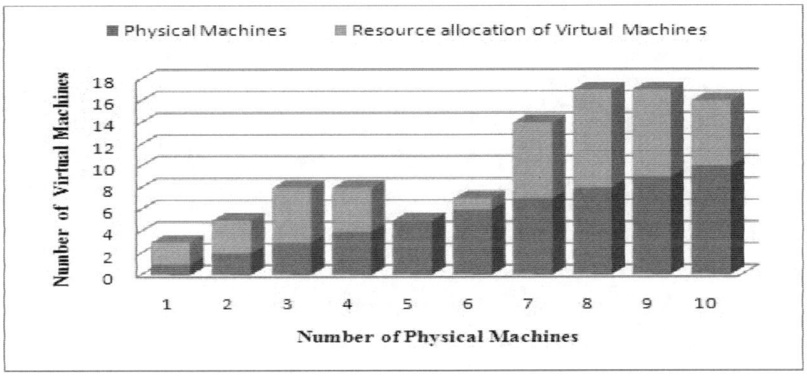

Figure 5.1: Resources allocation by VMBP placement algorithm

Scalability of cloud datacentres: In cloud data center scalability is the ability of PM working well when it is improved or changed in size. Therefore, VMBP algorithm with best fit strategy shows the placement of PM in VM has been changed the size of PM proves the efficient scalability of cloud data centres.

Comparison of benchmark functions fitness value for energy consumption: In VMBP algorithm the fitness value is calculated and compared with four different benchmark functions Sphere, Step, Graywang and Rastridge. The maximum number of iterations for benchmark functions are set to 100 because iterative procedure does not converge at all, the changes in the estimates from iteration to iteration do not get smaller and may even get larger. The fitness value is nothing but the distance travelled by each particle to form a swarm is calculated and compared with average value of benchmark functions. The Tab.3: shows the fitness value table for different benchmark functions.

Number of Iterations	***Fitness value***			
	Sphere function	***Step function***	***Graywang function***	***Rastridge Function***
10	39.64	62.48	62.52	54.38
20	39.64	50.20	44.27	62.70
30	39.64	52.44	50.20	62.80
40	48.05	48.05	50.27	50.27
50	39.92	58.37	70.29	39.92
60	48.05	62.57	50.20	50.20
70	62.46	70.59	50.27	58.38
80	62.48	39.88	62.48	62.48
90	52.29	70.79	62.68	58.15
100	48.05	50.39	52.44	62.72
Average	***48.22***	***56.57***	***55.56***	***56.20***

Table 3: Fitness value table for different benchmark functions

In Fig.5.1. VMBP algorithm fitness value for different benchmark functions showed graphically. We estimate the energy reduction in formation of each particle to make a swarm using the benchmark functions range. This is calculated by measuring the total energy consumption for all the particles to make a swarm. The minimum fitness value of sphere function is 39.64,step function is 39.88,Graywang is 44.27 and Rastridge is 39.92 therefore sphere function has got excellent performance of minimum fitness value in a given range. The maximum fitness value is also compared for all the benchmark functions to prove the highest point in a given range. The maximum fitness value of sphere function is 62.48, step function is 70.79, Graywang is 70.29 and Rastridge is 62.80 therefore step function has got excellent performance of in comparing maximum fitness value. Depends upon the average fitness value the sphere function travelled a less distance of 48.22, to make a swarm so sphere function has got the least energy consumption when compared with other three functions.

We can observe that in cloud data centres the placement of VM consumes energy will be better by using this VMBP algorithm with fitness value will achieve less energy consumption.

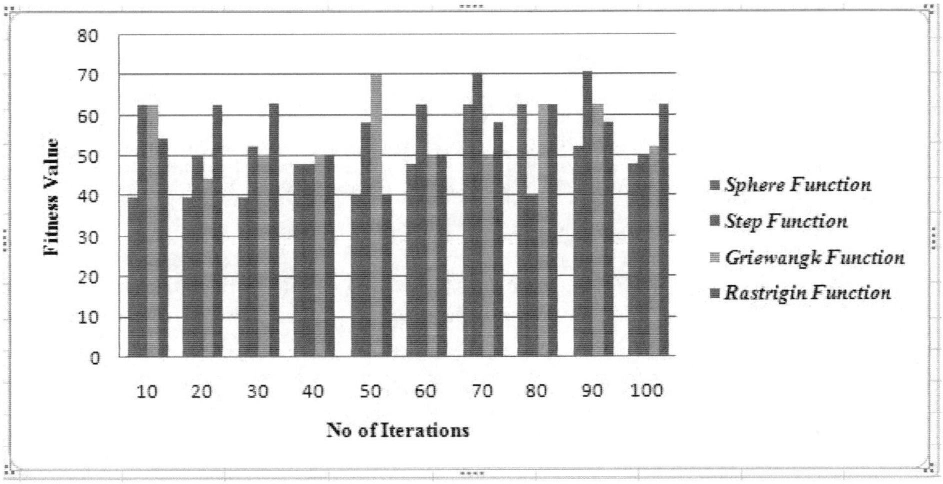

Figure 5.2: Fitness value of different benchmark functions

6. Conclusion and future work

In this paper the problem of dynamic virtual machine placement is expressed as a multi objective optimization tricky targeting to concurrently optimize maybe inconsistent purposes including a well-organized resource utilization, energy consumption and scalability of datacentres. In VMBP algorithm the particles integrate to make swarm and fitness value of swarm is calculated using the benchmark functions. The average fitness value of benchmark functions unimodal (Sphere, Step) and multimodal (Graywang and Rastridge) is compared and efficient energy consumption is obtained. The best fit strategy is used to get the efficient placement of virtual machines. Experimental result obtained shows VMBP algorithm has enhanced the resource utilization, reduces the energy consumption, and improves scalability of datacentres. Four benchmark functions are used to verify the VMBP algorithm for energy consumption. In the future, the projected algorithm will be verified on further benchmark functions for the projected effort.

References

[01] Priyanshu Srivastava and Rizwan Khan, *A Review paper on cloud computing*, International Journals of Advanced Research in Computer Science and Software Engineering ISSN:2277-128X (Volume-8, Issue-6), 2018.

[02] C.PandiSelvi and Dr.S.Sivakumar, *A review of virtual machine algorithm in cloud datacenter for server consolidation*, IJERCSE, Vol 5, issue 3, (2018),pp(182-188).

[03] Liang Hong and Ge Yufei, *GACA-VMP Virtual machine placement scheduling in cloud computing based genetic ant colony algorithm approach*, DOI:10.1109/UIC-ATC ScalCom-CBDCom-IoP.2015.

[04] C.PandiSelvi and Dr.S.Sivakumar, *Constraint programming approach based virtual machineplacement algorithm for server consolidation in cloud data center*, IJCSE, Vol 6, Special issue -8. (2018), E-ISSN: 2347-2693.

[05] Javaid, Nadeem, Sakeena Javaid, Wadood Abdul, Imran Ahmed, Ahmad Almogren,Atif Alamri, and Iftikhar Azim Niaz, *A hybrid genetic wind driven heuristic optimization algorithm for demand side management in smart grid*, Energies 10, no.3, (2017): 319.

[06] Cao, Zijian, Jin Lin, Can Wan, Yonghua Song, Yi Zhang, and Xiaohui Wang,*Optimal cloud computing resource allocation for demand side management in smart grid,* IEEE Transactions on Smart Grid 8, no. 4 (2017): 1943-1955.

[07] Vakilinia and Shahin, *Energy efficient temporal load aware resource allocation in cloud computing datacenters*, Journal of Cloud Computing 7, no. 1 (2018).

[08] Zhou, Ao, Shangguang Wang, Bo Cheng, Zibin Zheng, Fangchun Yang, Rong N.Chang, Michael R. Lyu, and Rajkumar Buyya, *Cloud service reliability enhancement via virtual machine placement optimization*, IEEE Transactions on Services Computing 10, no. 6 (2017): 902-913.

[09] Zahoor, Saman, Sakeena Javaid, Nadeem Javaid, Mahmood Ashraf, Farruh Ishmanov and Muhammad Afzal *Cloud Fog Based Smart Grid Model for Efficient Resource Management*, Sustainability 10, no. 6 (2018).

[10] Hao Jin, Deng Pan, Jing Xu, and Niki Pissinou, *Efficient VM Placement with Multiple Deterministic and Stochastic Resources in Data Centers*, 978-1-4673-0921-9/12/2012.

[11] Haojun Li, Zhongfeng Liu and Wanliang Wang, *Autonomous Groups Particle Swarm Optimization Algorithm Based On Exponential Decreasing Inertia Weight*, doi:10.21311/001.39.7.36, Rev. Téc. Ing. Univ. Zulia. Vol. 39, No 7, 289 - 296, 2016.

[12] Aisha Fathima, Nadeem Javid and Tanzeela Sultana, *Virtual machine placement via bin packing in cloud data centers*, MDPI Electronics 2018, 7, 389, DOI: 10:3390 / electronics / 7120389.

[13] Aisha Fathima, Nadeem Javid, Tanzeela Sultana and Mohammed Y Aalsalem, *An efficient virtual machine placement via bin packing in cloud data centers*, Research gate/ publication/330675892, (2019).

[14] Kumarasamy S,Mydhili and K Nair, *Bin Packing algorithms for virtual machine placement In cloud computing: A Review*, IJECE, Vol 9, No 1, ISSN 2088-8708, (2018), pp512-524.

[15] Mohamed Amine Kaouache and Sadok Bouamam, *Solving bin packing problem with hybrid genetic algorithm for VM placement in cloud*, ELSEVIER, Procedia Computer science 60 (2015) 1061-1069.

[16] Anil Dixit and Suchithra R, *Bin Packing Algorithm for Virtual Machine Placement in Cloud Data Centers for Resource Management*, International Journal for Research in Applied Science & Engineering Technology (IJRASET) ISSN: 2321-9653; Volume 6 Issue V, May 2018.

[17] Fale Mantim Innocent, Mwanret Alphonsus, Lar Nansel, Edwin Freedom Titus and Abdulmalik Dashe, *Best-Fit Virtual Machine Placement Algorithm for Load Balancing in a Cloud Computing Environment*, International Journal of Scientific & Engineering Research Volume 9, Issue 7, July-2018 1580 ISSN 2229-5518.

[18] Anton Beloglazov and Rajkumar Buyya,*Managing Overloaded Hosts for Dynamic Consolidation of Virtual Machines in Cloud Data Centers Under Quality of Service Constraints,* IEEE transactions on parallel and distributed systems, vol. X, no. X, 2012.

[19] Shawkat and Hamdan, *Testing of a modified particle swarm optimization algorithm using different benchmark functions*, DOI: 10.1109/SOFA.2009.5254866,July 2009.

[20] Alfonso Uriarte, Patricia Melin and Fevrier Valdez, *An Improved Particle Swarm Optimization Algorithm applied to Benchmark Functions,*2016 IEEE 8th International Conference on Intelligent Systems.

A Blockchain-based framework for Agriculture subsidy disbursement

Savita Bakare[1], Suneel C Shinde[2], Roopa Hubballi[3], Giridhar Hebbale[4], Dr.Vinayak Joshi[5]

[1]Department of Computer Science.&Engg, KLE Dr. M.S. Sheshgiri College of Engineering and Technology, Udyambhag, Belagavi -590008, Karnataka, India

[2]Department of Master of Computer Applications, KLE Dr. M.S. Sheshgiri College of Engineering and Technology, Udyambhag, Belagavi -590008, Karnataka, India

[3]Department of Computer Science.&Engg, KLE Dr. M.S. Sheshgiri College of Engineering and Technology, Udyambhag, Belagavi -590008, Karnataka, India

[4]Department of Biomedical Engineering, KLE Dr. M.S. Sheshgiri College of Engineering and Technology, Udyambhag, Belagavi -590008, Karnataka, India

[5]Department of Computer Science.&Engg, Angadi Institute of Technology and Management, Savagaon Road, Belagavi-590 006, Karnataka, India

Email: *savita.bakare@klescet.ac.in

Abstract - Agriculture is the primary livelihood factor in the growth of India's economy. One of the significant challenges faced by the Government of India in the agriculture sector is to increase the delivery of subsidies by re-engineering the existing process. One of the solutions initiated by the government of India on 1st January 2013 with the highest priority is called a Direct Cash transfer scheme under the Direct Benefit Transfer (DBT) scheme. It involves crediting subsidies (cash/payments) directly to the farmer's account. However, the major problem associated with this scheme is that the government subsidies do not reach all the eligible farmers due to corruption, delay, intermediaries, and other reasons. The present direct cash transfers system's major drawbacks such as (key loopholes are) lack of proper auditing, inability to track subsidy, and inadequate digital banking infrastructure especially in rural areas. In this regard, the disbursement of direct cash transfer scheme is addressed benefitting the farmers directly via Blockchain Smart Contracts technology thereby ensuring transparency, de-duplication, reduced delays, and reducing instances of fraud in the existing government subsidy delivery system. We aim to develop a blockchain-based smart contracts prototype model called a single-window approach for automating farmer's subsidies and eliminating middlemen. The automated model will be fast, transparent, and immutable by reducing the complexities faced by Indian farmers.

Keywords: **smart contracts, BT scheme, direct cash transfer, blockchain systems, e-governance, distributed ledger technology, disbursement**

1. Introduction

Background

In India's economy, the agriculture sector is ranked fifth in production, consumption, export, and expected growth in the FY2020-21. Agriculture is the primary livelihood for more than 70% of the rural population and employs over 60% of the total population of India. In the agriculture sector, one of the challenges faced by the Government of India is to increase the quick delivery of agricultural subsidies and ensures accurate targeting of the beneficiaries. In this regard, in January 2013 the Government of India has launched the Direct Cash Transfer scheme under DBT for fund distribution and delivery system. In this scheme, the subsidies (cash/payments) are directly credited to the farmer's account and validated by the Unique Identification Authority of India (UIDAI) to bring transparency, reduce fraud and delays happening by the hierarchy of government administrative offices.

The DBT schemes are of three types :(a) Cash Benefit (Government transfer cash directly to farmer account called as subsidy like PAHAL, NSAP, MGNREGA, etc.), (b) In-Kind Benefit (Government transfer free cash directly to farmer account through an intermediate agency like Public Distribution System (PDS), Food Corporation of India (FCI), mid-day meals, etc.) and (c) Mix Benefit (Combination of Cash benefit and Kind benefit). As per the survey report, 450 DBT schemes under 56 ministries have been launched in India and are delivered to more than 900 million beneficiaries. In which 97 are cash schemes and 63 are In-kind schemes, while the remaining are a mix of cash-kind schemes. However, the major problem is that these government subsidies often do not reach all the eligible farmers due to corruption, delay, intermediaries, and other reasons.

The government record shows only 1% of the cash transactions have failed, which means 99% is a success, which is not valid. Only 20% of the farmer's complaint about not getting subsidies. This problem occurs because the National Payment Corporation of India links the details of the latest bank accounts with the Aadhaar to DBT payments. So, farmers verify the bank account, which was initially receiving the subsidy, and finds that the money has not been deposited. The present DBT based direct cash delivery system's key drawbacks are lack of proper auditing, inability to track subsidy, and inadequate digital banking infrastructure in rural areas. In this regard, to increase the transparency in the government subsidy delivery system for farmers, the Indian states and central governments have been actively looking for technological up-gradation into agri-practices.

Thus, the combination of DBT based direct cash transfer scheme by adopting Blockchain-based Smart Contracts technology will improve the disbursement of existing subsidy/welfare benefits to the farmers. It can target subsidies for food grains, seeds, fertilizer, and fuel etc., to know the availability, consumption, and delivery mechanism for bringing out inefficiency, accountability, and transparency. In simple terms, Blockchain is a systematic-immutable backtracked linked-list database to store and access value-transactions that have happened and may happen in the future directly without any intermediaries. It can significantly improve transparency, reduce corruption, tamper-proof, and enhance security by reducing the overhead costs of auditing, accounting, and legal issues.

In the direct cash transfer scheme, insight into who is being issued subsidies, what they are being used for, and each payment's wisdom should be made available for farmers by the government. In this paper, we address the solution to the above problems by adopting blockchain-based smart-contracts technology. In this system, the existing farmer's database will be linked to a blockchain database (smart contracts) for providing automation subsidies between the government and the farmers. It makes the system more transparent, efficient, and trustworthy without charging transaction fees and removes intermediaries who may misuse/divert subsidy. Intelligent contracts (or smart contracts) are the digital agreement to automate the transactions when the agreement parties meet the agreement terms and conditions.

2. Motivation and Objective

The primary motivation is the disbursement of direct cash subsidy benefit under the DBT scheme to all the Indian farmers without any intermediate agencies and delays. To achieve this, the objective is to improve and resolve the above said key loopholes of the existing government's fund delivery system. The solution converses on a Blockchain-based smart contracts prototype known as the single-window approach. The automated model ensures to be fast, transparent, and immutable by reducing the complexities faced by the Indian farmers. It improves the system's security while considering many aspects such as overhead expenses of auditing, accounting, and legal issues. The identified complexities for subsidy transfer under the existing DBT system and its solution conversing on blockchain-smart-contracts technology can be addressed as:

2.1 *Mandatory enrollment:* The new DBT policy requires farmers to link an Aadhaar number with their bank account. However, the majority of farmers do not have either of them. If enrollment is not mandatory, then the verification systems (Identity card and Aadhaar number) must coexist. Still, the farmers with multiple cards will choose to opt-out of the process of the Aadhaar system, and also, it cannot address problems related to cash/payments.

2.2 *Removing Intermediate/Middlemen:* The subsidy is often misused by the intermediate persons or middlemen who divert the sanctioned subsidy/fund from the farmer accounts to their accounts, and this has to be eliminated by adopting a blockchain-based system.

2.3 *Maintaining privacy:* Data collected while issuing Aadhaar may be misused. India does not have a specific law governing issues related to privacy. The authority stores the details of every request for authentication, which provides insights into the usage patterns of an Aadhaar number holder which may be misused. Hence, there is a need to secure and track this data by applying the concepts of blockchain smart contracts.

2.4 *Transparency in subsidy-chain amounts:* The disbursement of subsidy amounts in the supply-chain may be misused. The eligible and needy farmer will not avail of the subsidy benefit entirely. Using blockchain smart contracts, we can track the subsidy amount.

2.5 *Farmers' digital Identity Repository:* It is used to store, share, and validate personal bio-data of farmers by the use of cryptographic methods to increase the trust via smart contracts.

2.6 *Removing Commission:* Bank charges for every transaction, and also this process is time-consuming. To avoid this, blockchain-based smart contracts ensures that there is no transaction fee charged.

3. Literature Review

This section reviews and highlights the related work to know which agri-digital domains have been explored in Blockchain for solving the above-stated problems. In recent research [1] the authors outline the blockchain technology implementation in government projects and had proposed various initiatives to integrate stakeholders into food production and supply chain. For technology benefits, the blockchain barriers and challenges can be addressed by the government research fund efforts to foster digitization. This is possible by educating farmers through various training methods. In [2] the open cyber security challenges of smart farming, security and privacy issues, use cases w.r.t cloud and edge computing, and relevant research work in the cyber security domain were addressed. The authors [3] had discussed blockchain applications, problems, and solutions related to agricultural and food sectors like agricultural insurance, smart farming, food supply chain, and farm products transactions. They also addressed uploading the existing data into the Blockchain database may be very costly. The various data security issues [4] such as confidentiality, integrity, availability, and implementation solutions were possible using a private blockchain-smart-contract in agriculture. For privacy, the solution is a private blockchain network; for integrity, the solution uses a hashed-chain structure for verification over blockchain network and for availability, the solution is to store the transaction data (hashes) in the tamper-proof smart contracts. In another work [5] the authors had discussed sustainable agricultural data management in rural areas for energy consumption, data planning for making decisions w.r.t data collection, storage, and performance to determine the reliability factor of blockchain. In [6] the current research trends, applications, and challenges of smart contracts based on blockchain technology were discussed.

4. Problem Statement

The DBT based direct cash schemes have various loopholes such as the involvement of intermediaries, security & privacy issues as well commissions charged by the various banks. The DBT scheme implementation in the [7] Jharkhand state of India has put a question mark on the efficiency of the system as the nexus between banks personnel, intermediaries, government and school authorities seem to have siphoned off significant portions of the funds, gives an excellent example of the flaws present in the system. As in the Jharkhand case, the authorities and intermediaries, including banking correspondents, are alleged to have obtained the relevant documents and particulars, including fingerprints, genuine and fake beneficiaries, certified their authenticity, and perpetrated a scam. According to a report submitted by Jameel Poverty Action Lab, the loopholes of the DBT scheme and ineptitude of payee to acquire the benefits has hindered the faith in the current system [10]. The critical issues identified in the existing system are:

a) Over dependency on banking systems.
b) Erroneous amount of fund transfer.
c) No audit/grievance redressal mechanism.
d) Middlemen Commission
e) Fake beneficiaries

To address the above-said issues, we use a public-permissioned blockchain. The reason is a public-permissioned blockchain network, infuses the space by linking the public-permission less networks i.e., Bitcoin/Ethereum with the private-consortium network to attain the prime quality of distributed models. Transparency is achieved through access control and consensus mechanisms by maintaining transaction sequences [11]. Moreover, the issues related to the performance of the public-permissioned blockchain [12] are ranked high to overcome delay and achieve competence. In the long run, this system highlights privacy more than public-blockchains [13]. By considering these issues we formulate the problem statements as:

a) Design and develop a secured system based on blockchain to track subsidy/fund disbursement for the enrolled farmer (beneficiary).
b) This system has to be adequately auditable and fully automated.
c) The designed system must be secured and should provide privacy and paramountcy of data on the Blockchain.
d) The system must remove the fake beneficiaries and the commissions charged by the bank.
e) Design and develop a robust algorithm to calculate a unique blockchain id/address and OTP-based on the Aadhaar number.

5. Existing System

The existing system has various complexities such as duplication and tampering of the submitted farmland records, fraud cash transfer in the bank accounts, and delay in the process which is faced by the farmers as depicted in figure 5.1. They need to register with the various government departments such as the land Records Department, the Government of India Treasury Department, customer bank to claim the cash benefits provided by the government fund delivery system.

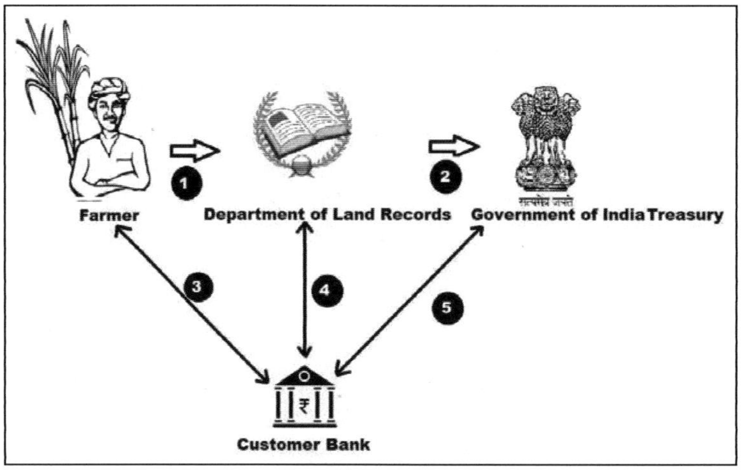

Figure 5.1. Existing system

The existing process is summarized in five steps as follows:

1. In the first step, the farmers need to register with the Land Records Department by providing farmland records.
2. In this step, the Land Records department will submit the details of the farmer's farmland to the Government of India Treasury Department. The department verifies the details and allots cash benefit to the individual farmer.
3. The farmer has to open an account in a private or public bank validated by UIDAI, such as an Aadhar card.
4. In this step, the bank verifies both farmer's details submitted at Land Record sources and facts presented at the bank and then links Aadhar card with the farmer's bank account.

5. The government of India's own Treasury Department verifies the details from banks and the Land Records office and initiate direct cash transfer to the farmer's bank account.

6. Proposed System

The proposed system reduces the farmers' complexities and problems as it ensures to provide flexibility in terms of a single-window blockchain approach to furnish security and privacy. The proposed system is as shown in figure 2, which is auditable, transparent, and automated. Here we summarize the entire process of the proposed system as follows:

1. In this step, the farmer will register to the blockchain-based fund transfer system by uploading his farmland records and get authenticated via UIDAI Aadhar card. In turn, the blockchain-based system will generate a unique digital identity such as wallet address using robust cryptographic calculations.
2. In the second step, the three significant entities of this system will access the details provided by the farmer, the Banks and the Department of Land Records will verify the details mutually such as farmland records, Aadhar number, mobile number, etc.,

After verifying the Blockchain-based fund transfer system's details, the Government of India will initiate the fund transfer process. So in this way, the system removes the middlemen, fake/fraudulent identity entities and provides immediate transfer of funds concept and the commission charges imposed by various banks.

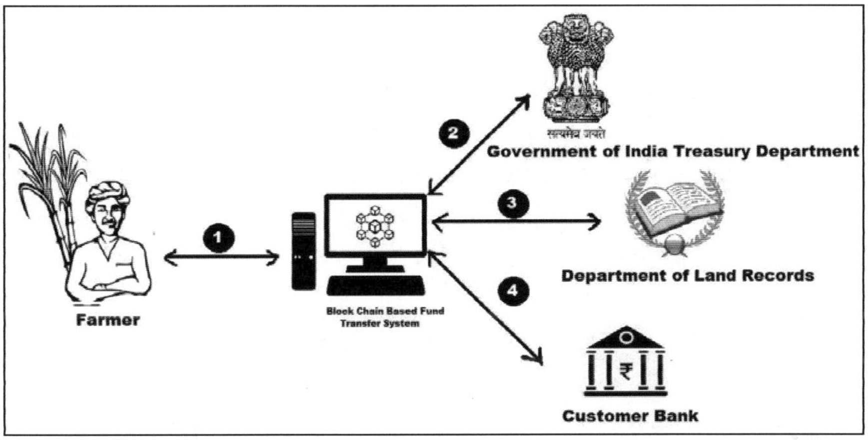

Figure 6.1. The overview of Blockchain-based Fund Transfer System

In figure 6.1, It depicts the blockchain-based fund transfer system using a public-permissioned blockchain network. A public-permissioned block-chain network is a new type of network which fills the gap between the public and private consortium networks [8] to infuse the space by a distributed model. The GUI of the proposed system is designed using AngularJS to communicate across the systems. We have used the NodeJS Express server to build a blockchain system using hyper ledger fabric and composer [14] [16]. The four users of the system are the Farmer, Government Personnel (Treasury Department), Department of Land Records, and Customer Bank. We propose various assets for the systems, such as Aadhar card number, OTP, and subsidy amount. We also propose to write several fund transactions, which is based on Smart Contracts.

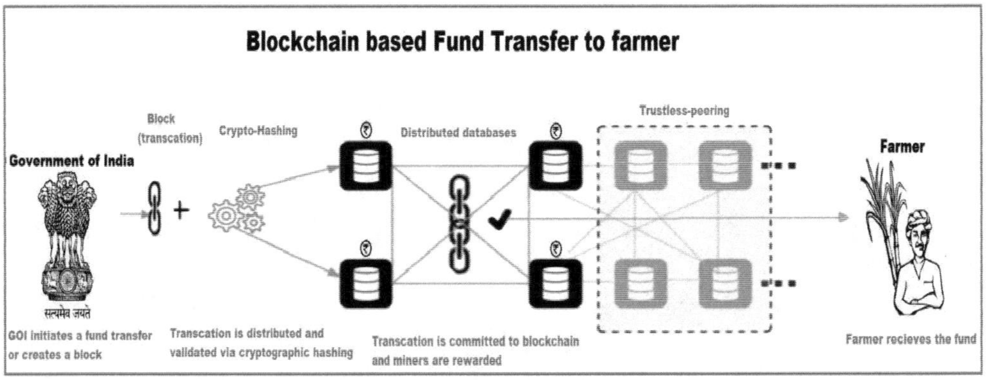

Figure 6.2. The architecture of Blockchain-based fund transfer System

The architecture shown in figure 6.2, has three major components which provide transparency, security, and sanctity summarized as:

i. Primarily, the system will create blocks in the blockchain to store and link the transaction data. Each block stores the previous block hash data for cryptographic security, and after that, crypto hashing is done using some built-in algorithms such asMD5 (message digest 5) SHA-1 (secure hash algorithm). The output of MD5 is 128-bit digest and SHA-1 is 160-bits digest. A cryptographic hash function is a particular class of hash functions sometimes called 'digest'. It is a kind of 'signature' for a text or a data file with various properties, making it ideal for cryptography.
ii. In the second step, the system stores the transaction details in a distributed database. Block-chain's distributed ledger is an immutable and secured digital public ledger continuously growing the cryptographically secured database.
iii. In trustless peering, the trust is the other name given to the mathematical evidence for verification of nodes in the peer-to-peer network. The nodes performing transactions in the blockchain peer-to-peer network need not necessarily know the real identity of other node is called as blockchains trustless property [11][15].

7. Results and Discussion

In this section, we discuss the proposed blockchain-based fund transfer prototype model, its scope, and functionalities. Before using this system, we assume that the farmer has the following documents:

1) Farmer has an Aadhaar Card/ID,
2) Farmland records attested/verified by the department of Land records,
3) We believe that farmer has a mobile phone to receive the generated OTP and
4) We also think that farmers can use the designed portal/application.

Prototype

In figure 7.1, we illustrate the login information for the blockchain-based prototype model. The farmer has to input his Aadhar card number and OTP on his mobile. The entered information is stored in the blockchain database by creating blocks in the blockchain system and making smart contracts. This system has four users: Government Authority, Farmer, Department of Land Records, and Banks. As Government Treasury department is having a privilege to verify subsidy/fund transfers to the beneficiaries/farmers, deliver subsidy, and generate a report via smart queries.

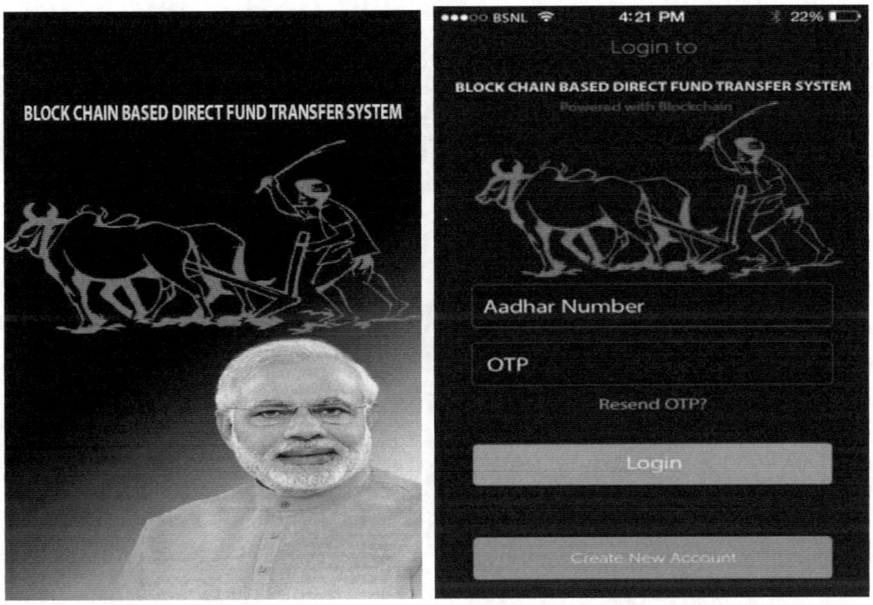

Figure 7.1. Login screen

As shown In figure 7.2, The farmer can upload the farmland documents in either of the formats such as JPEG or PDF. In turn, the land record department will verify the farmer's records/documents by the block-chain based system using participants of system like miners who will in turn submit a report to the government of India.

Figure 7.2. Upload screen

The treasury department then initiates the transfer of the subsidized amount to be credited to the individual farmer.

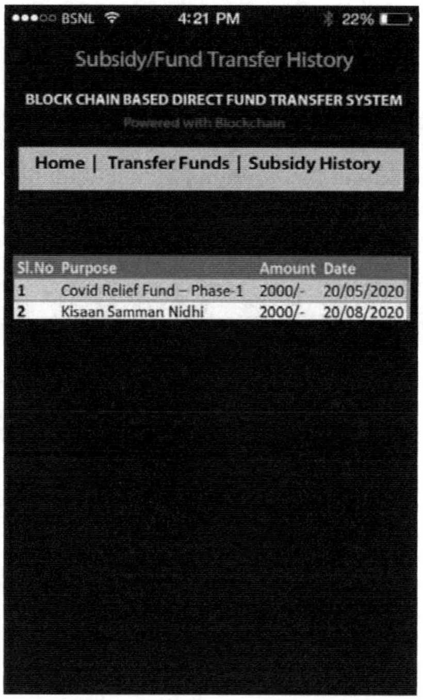

Figure 7.3. Transfer Records

Figure 7.4. Fund transfer initiate screen

In the above figures 7.3 and 7.4, the other participants of blockchain-based fund transfer system as Government authority and Department of Land records can view the subsidy or fund transfer history and based on the document submitted to the system been verified thoroughly by the Department of Land records are eligible for fund transfer. The GOI initiates the fund transfer to the eligible beneficiary/farmer to his Blockchain wallet ID.

8. Conclusion

We have discussed the significant challenges and the critical loopholes faced by the Government of India in the existing DBT cash transfer scheme. The solution converses on blockchain-based smart contracts to improve the existing government's fund delivery system. In this paper, we have discussed and designed a prototype model called a single-window approach based on blockchain technology to reduce the farmer's complexities and major issues faced by them in India. The automated model ensures to provide security and privacy, auditability as well as transparency and immutability. Thus, the direct cash delivery prototype can be adapted and implemented by using Blockchain-based technology with several tactile benefits. In the end, we discussed the functionalities of the robust system developed using Blockchain-smart-contracts.

IOP Conf. Series: Materials Science and Engineering 1110 (2021) 012008 doi:10.1088/1757-899X/1110/1/012008

Acknowledgement

In this research work, all the authors have contributed equally and are thankful to KLE Dr. M S Sheshgiri College of Engineering and Technology, Belagavi and Angadi Institute of Technology and Management, Belagavi, Karnataka, India for supporting and providing facilities to carry out research work.

Reference

[1] Andreas K et al 2018*J.Trends in Food Science & Technology* **91** 640-652
[2] M. Gupta et al 2020, Security and Privacy in Smart Farming: Challenges and Opportunities, *J. IEEE Access* **8** 34564-34584
[3] H L Gururaj, A Manoj Athreya, Ashwin A Kumar, Abhishek M Holla, S M Nagarajath, V Ravi Kumar "*A New Era of Technology*", Cryptocurrencies and Blockchain Technology Applications, May 2020, Wiley Digital Library https://doi.org/10.1002/9781119621201.ch1
[4] Xiong et al 2020 Blockchain Technology for Agriculture: Applications and Rationale *J Front. Blockchain* **3** A7
[5] Li X, Wang D, Li M2020 *J.Cleaner Production* **271** 2503
[6] L. Song, X. Wang and N. Merveille 2020, *Conf. TEMSCON, USA* 1-5
[7] Daniel Macrinici et al2018 *J. Telematics and Informatics*35 2337-2354
[8] Financial Express 2020 Learn from Jharkhand scam; govt schemes need independent ground-level auditing *https://bit.ly/3hvOkCp* **Nov 3**
[9] James Carlyle 2020 How "public-permissioned" blockchains are not an oxymoron. *https://bit.ly/3hDF67q* **02-2020**
[10] Fakharul Hassan et al 2020 *J. IEEE COMST* V2 1904.00733
[11] Joe C. Mathew 2017 All Cost, No Benefit *https://bit.ly/394ap7B* **03-2020**
[12] D. Mingxiao et al 2017 *Proc. Int. Conf on Systems, Man, and Cybernetics* 2567-2572
[13] Z. Zheng et al, 2017 *Proc. Int.Congress on Big Data (Big Data Congress)* 557-564.
[14] E. Androulaki et al 2018 *Proc. Int. Conf on the Thirteenth EuroSys* **30**
[15] Dr. Mohammad Nauman 2018 *Packt Publishing*
[16] H L Gururaj, A Manoj Athreya, Ashwin A Kumar,S M Nagarajath, V Ravi Kumar *"Adoption of pets in distributed network using blockchain technology",* August 2020, International Journal of Blockchains and Cryptocurrencies, Inderscience Publishers, https://doi.org/10.1504/IJBC.2020.108996

A Review of Different Approaches for Detecting Emotion from Text

Ashritha R Murthy[1], Anil Kumar K M[2,]

Assistant Professor[1], Department of Computer Science and Engineering, Sri Jayachamarajendra College of Engineering, JSS Science and Technology University.
Associate Professor[2], Department of Computer Science and Engineering, Sri Jayachamarajendra College of Engineering, JSS Science and Technology University.

E-mail id: ashrithar.murthy@sjce.ac.in

Abstract. Emotion detection and analysis is one of the challenging and emerging issues in the field of natural language processing (NLP). Detecting an individual's emotional state from textual data is an active area of study, along with identifying emotions from facial and audio records. The study of emotions can benefit from many applications in various fields, including neuroscience, data mining, psychology, human-computer interaction, e-learning, information filtering systems and cognitive science. The rich source of text available in the Social media, blogs, customer review, news articles can be a useful resource to explore various insights in text mining, including emotions. The purpose of this study is to provide a survey of existing approaches, models, datasets, lexicons, metrics and their limitations in the detection of emotions from the text useful for researchers in carrying out emotion detection activities.

Keywords: Emotion, Lexicon, Machine learning, Psychology, Deep learning

1. Introduction

The emotion is an interdisciplinary field involving psychology, computer science and others. In psychology, emotions are expressed as psychological state differently connected with contemplations, sentiments, behavioural responses, and a level of delight or displeasure [1]. In computer science, emotions can be identified in the form of audio records, video records and text documents. Analysing emotions from the text documents seems to be challenging due to the fact that textual expressions are not always directly use the emotion related words, but often outcome from the understanding of the meaning of concepts and interaction of concepts mentioned in the text document.

Emotion expressions are the crucial form of communication in interpersonal relationship [2]. It can be expressed into positive emotion, negative emotion or neutral [3]. In general the positive emotions are expressed as *happy, excited, joy, pride* and negative emotions such as *sad, disgust, fear, depressed* and so on. In such a way, the emotions are expressed in various forms to communicate and the rich source of textual information is obtained from the social networking sites such as YouTube, Twitter, and Facebook etc., where people are spending most of their time in posting and expressing their emotions [4]. By considering the textual data available on the blogs, it is helpful to identify the intensity of emotion of an individual. For example "Really happy with this purchase" [5] express the positive emotions from a customer about the purchase of product. The term "Really" intensifies the emotion expressed by the customer. In this case it implies it is more positive emotions. Another customer review on the same product [5] expressed negative emotions on the purchase of the product such as "Really disappointed. Alexa has to be plug-in to wall socket all the time. My fault is for not checking". Here the customer intensifies more negative comment about the purchase. By considering the intensity of emotion through the text, it helps to predict individual emotions. Also, it helps to know the state of emotions of the person that can assist friends and family to take preventive measures against accidents or self harm. The contribution of the paper is to provide the state of the art approaches used in the detection of emotion from text.

In section 2, the relevant emotion models are discussed. In section 3, the existing resources such as the Corpora and lexicon are discussed. In section 4, computational approaches used in literature are discussed. Evaluation metrics used in literature are discussed in section 5. Summary of the existing approaches are discussed in section 6. Finally conclusion is discussed in section 7.

2. Emotion Models

Emotions are recognized by humans and this effect has influenced the way emotions are viewed in scientific terms. Researchers in psychological science believe that individuals have internal mechanisms for a limited collection of responses (usually happy, sad, anger, disgust, and fear) that can be assessed in a simple and objective manner once activated [6]. In order to represent emotions, the three major approaches are used in the psychology research [7] are

1.1 *Categorical approach*: This approach of emotion involves placing emotions into categories or into distinct classes that are basic and universally recognized [7].The emotions are independent and also depend on how an experience perceiving the situation which can be categorized into six basic fundamental emotions i.e., *{happiness, fear, sadness, surprise, disgust and anger}* by Paul Ekman[8].

The model by Robert Plutchik[9] proposed eight fundamental emotions, i.e. acceptance / trust and anticipation. The 8(eight) emotions are expressed in opposite pairs as *surprise verus anticipation, joy versus sadness, anger versus fear and trust versus disgus*t. There are varying degrees of intensity for each emotion, according to Plutchik, which occurred as a result of the perception of events via an experiencer.

The model of Orthony, Clore, and Collins (OCC) [10] differed from the "basic emotions" analogy proposed by Ekman[8] and Plutchik[9]. They accepted, however that emotions emerged as a result of how the events and emotions perceived by an individual differed based on the degree of intensity. Emotions have been categorised into 22, 16 emotions were added to the basic emotions proposed by Ekman[8] and a much wider representation of emotions as {*envy, relief, appreciation, self-reproach, shame, reproach, pity, admiration, disappointment, grief, gratification, fears-confirmed, gloating, hope, like and dislike*}.

1.2 *Dimensional approach:* In this approach, it considers emotional states to be bound to each other rather than to be independent. Hence, it is represented in dimensional space [7] (uni dimensional and multi-dimensional) describing how emotions are connected based on the event and their degree (low to high) of occurrence. This article explores more on multi-dimensional models for emotional representation.

- Russell's circumplex model [11] represent the emotions in two dimensional model. As emotions are not independent it distinguished as Arousal (Activation and Deactivation) and Valence (Pleasantness and Unpleasantness) where the dimension Arousal refers to how excited or apathetic an emotion is and the dimension Valence refer to how positive emotion and negative emotion is [7]. Figure 2.1 shows Russell's circumplex model of affect.

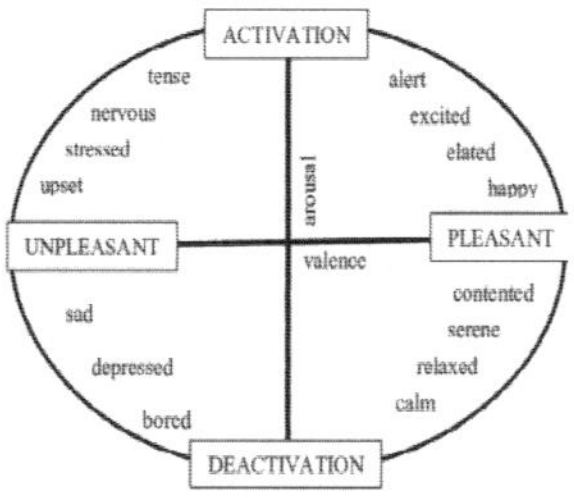

Figure 2.1: Circumplex model by Russell [11].

- In the two-dimensional model, Plutchik's wheel of emotions [9] represents emotions as a wheel of emotions, as seen in figure 2.2. In the concentric circle, the wheel represents emotions, with the inner core emotions being variants of the eight basic emotions, then the eight basic emotions in the outermost areas of the wheel, and finally combinations of the primary emotions. The wheel shows how the emotions are related, depends on the location on the wheel.

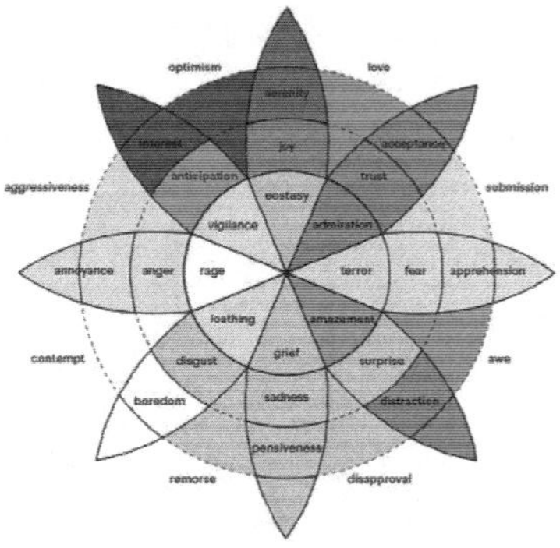

Figure 2.2: Plutchik's Wheel of emotions [9]

- Russell and Mehrabian [12] represent emotions into three dimension model such as pleasantness (or positiveness), arousal (or responsiveness),and potency (or dominance). In 2D representation the emotions are distinguished as Arousal (Activation and Deactivation) and Valence (Pleasantness and Unpleasantness) and the third dimension
as Dominance Power [2]) refers to the degree to which an experienced had been control of their emotions. Figure 2.3 represents 3D emotion space.

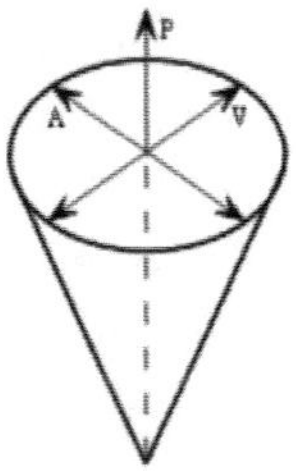

Figure 2.3: 3D Emotion Space (Valence, Arousal, and Power) Images adapted from [7].

- Parrot [13] organized emotion into three hierarchical structures namely primary, secondary and tertiary emotions with joy, anger, fear, love and surprise as primary set of emotion. Table 1 represents different emotions used in Parrot Emotion Taxonomy.

Table1: Parrot's Emotion Taxonomy

Primary emotions	Secondary emotions	Tertiary emotions
Love	Affection	Caring, sentimentality, adoration, fondness, compassion, tenderness, love, attraction, affection, liking.
	Lust	Desire, arousal, passion, Infatuation, lust.
Joy	Cheerfulness	Euphoria, satisfaction, gladness, ,bliss, happiness, Jubilation, delight, elation, enjoyment, joviality, joy, Amusement ecstasy, jolliness, gaiety, glee, cheerfulness.
	Zest	Thrill, excitement, enthusiasm, exhilaration. zeal, zest.
	Contentment	Pleasure, Contentment.
	Pride	Triumph, Pride.
	Optimism	Optimism, hope, Eagerness.
Surprise	Surprise	Astonishment, surprise, amazement.
	Irritation	Agitation, annoyance, irritation, grouchiness aggravation,
Sadness	Sadness	Gloom, grief, melancholy despair, glumness, misery, unhappiness, hopelessness, sadness, depression, sorrow, woe,
	Disappointment	Dismay, displeasure , disappointment
	Shame	Regret, shame, remorse, guilt
Anger	Exasperation	Frustration, exasperation,
	Rage	Bitterness, loathing, wrath, dislike, hostility, spite, resentment rage, ferocity, hate, scorn, fury, anger, vengefulness, outrage.
	Disgust	Contempt, revulsion, disgust.
	Suffering	Hurt, agony, anguish suffering,

Fear	Horror	Horror, panic, hysteria, alarm, fear, terror, shock, mortification fright.
	Nervousness	Apprehension, tenseness, anxiety, dreads worry, uneasiness, distress, nervousness.

1.3 *Appraisal based approach:* The Dimensional Model can be seen as an extension of this approach. It contains componential emotion models based on the theory of appraisal. Appraisal theory describes how different emotions, in various participants and on different times, can arise from the same event [14]. The emotions can see changes through the significant components such as cognition, expressions, physiology, motivation, motor, reactions, and feelings [7].

In the categorical approach, emotional states are restricted to a limited number of distinct types and it can be difficult to resolve a complex emotional situation or mixed emotions**.** In such cases, dimensional approach can be used, however not all basic emotions fit into it. Hence the advantage of the componential model [7] can be used according to the variability of different emotional state due appraisal pattern.

3. Resources

In this section we discuss the Corpora and lexicons used in research works related to the detection of emotions using text.

3.1 Corpora

The Corpora is the collection of linguistic data that are used to detect the emotions from text. Table 2 describes various corpuses available in the literature.

Table2: List of Corpora

Dataset	Features	Downloadable Link
Aman [15]	Blog post.	http://saimacs.github.io/ Available on request
Alm [16]	185 Children stories	http://people.rc.rit.edu/~coagla/affectdata/index.html
ISEAR[17,18]	International survey on antecedents and reaction	http://www.affective-sciences.org/index.php/download_file/view/395/296.
	Cross cultural studies	https://www.kaggle.com/shrivastava/isears-dataset
SemEval 2007[19]	News headlines	http://web.eecs.umich.edu/~mihalcea/affectivetext/#resources.
SemEval 2017[20]	Social media and news	http://alt.qcri.org/semeval2017/task4/index.php?id=download-thefull-training-data-for-semeval-2017-task-4
SemEval 2018[21]	Tweets	https://competitions.codalab.org/competitions/17751.
SemEval 2019[22]	Textual dialog betwwen two individuals	https://www.linkedin.com/groups/12133338/.
EmoBank [23]	Blogs, newsheadline, newspapers, letters and	http://saifmohammad.com/WebPages/EmotionIntensity-SharedTask.html

	travel guides	
CrowdFlower [24]	Tweets	https://www.crowdflower.com/wp-content/uploads/2016/07/text_emotion.csv
WASSA-2017 Emotion Intensities(EmoInt) [25]	Tweets	http://saifmohammad.com/WebPages/EmotionIntensity-SharedTask.html
Cecilia Ovesdotter Alm's Affect data [26]	Stories	http://people.rc.rit.edu/~coagla/affectdata/index.html
DailyDialog [27]	Dialogues	https://www.aclweb.org/anthology/I17-1099/
Emotion Stimulus[28]	Designed out of FrameNets	http://www.site.uottawa.ca/~diana/resources/emotion_stimulus_data
MELD data [29]	Dialogues from Television show	https://github.com/SenticNet/MELD
SMILE dataset[30]	Tweets	https://figshare.com/articles/smile_annotations_final_csv/3187909
Dens Dataset[31]	Information obtained from wattpad web histories and project Gutenberg literature	Available on Request
The Valence and Arousal dataset[32]	Posts of FB (Facebook)	http://wwbp.org/downloads/public_data/dataset-fb-valence-arousalanon.csv
Grounded emotions[33]	Tweets	http://web.eecs.umich.edu/~mihalcea/downloads.html$\delimiter"005C317$#GroundedEmotions
Emotion Lines	FB message chats and Television show {FRIENDS}	https://sites.google.com/view/emotionx2019/datasets; Available upon registration
Amazon Alexa Reviews emotions[5]	Alexa Reviews	https://www.kaggle.com/sid321axn/amazon-alexa-reviews

3.2 Lexicons

The lexicons used to detect the emotions from the text that are available in literature and are discussed in Table 3.

Table 3: List of lexicons

Lexicon	Description	Link
AFINN [35]	It is manually rated word for valence within an integer between minus five(-5) and plus five(+5).	https://github.com/fnielsen/afinn.
Sentiment140 Lexicon [36]	It is automatically generated from tweets that contain emoticons.	http://saifmohammad.com/WebPages/lexicons.html#NRCTwitter
NRC Hashtag Emotion Lexicon [37,38]	It is created automatically from tweets that contain hashtags with emotional terms such as #happy,	http://www.saifmohammad.com/WebPages/AffectIntensity.htm.

	#sad. It is primarily associated with words such as anger, fear, anticipation, surprise, happiness and trust.	
Bing Liu Lexicon [39]	This Lexicon consists of a list of positive and negative opinion words or sentiment words.	https://www.cs.uic.edu/~liub/FBS/sentiment-analysis.html#lexicon.
NRC Hashtag Sentiment Lexicon [40]	It is automatically generated from tweets that include sentiment-word hashtag such as #amazing. It is associated with positive or negative sentiment.	http://saifmohammad.com/WebPages/lexicons.html#NRCTwitter.
WordNet [41]	Online English lexical Database. It consists of verbs, nouns, adjectives and adverbs into a set of synonyms called synsets.	https://wordnet.princeton.edu.
NRC Word-Emotion Association Lexicon [42,43]	Use Amazon's Mechanical Truck, Lexicon Annotated Manually. Consist of 8 emotions and two positive and negative sentiments are included	http://saifmohammad.com/WebPages/NRC-Emotion-Lexicon.htm.
EmoSenticNet [44]	Emotions labels to Sentic Concept	https://www.gelbukh.com/emosenticnet/

4. Computational approaches:

The different approaches proposed in the literature for the identification of emotions from the text were discussed in the following sections.

4.1 Keyword based approach:

In this approach, it exploits the knowledge of key features that are combination with emotion labels using a lexicon such as Word-Net Affect and SentiwordNet, linguistic rules are applied and sentence structures are exploited. Further text preprocessing has to be performed on the given dataset which includes stopword removal, tokenization and lemmatization. In addition, keyword spotting and emotion intensity are evaluated including with Negation checks. Finally, it determines the emotion label for each sentence.

CC Liu et al.[14] express this approach is based on the set of keyword which contains emotions. Without a keyword in a sentence means that it doesn't contain any emotions in them. For example: "Today, I passed my exam with distinction" and the sentence as "Hurray! Today, I passed my exam with distinction" "I passed my examination with distinction today" could indicate the same emotion (joy), but if "hurray" is the only keyword to detect this emotion, the former without "hurray" might remain undetected. They introduced an architecture aimed at providing diverse contexts with a systematic understanding of textual input and better flexibility. That is, with semantic analysis, the extraction of semantic information, the design of ontology based on emotion models and the adoption of new keywords with case-based reasoning.

Sailunaz, Kashfia, and Reda Alhajj [45] used the latter data to create generalized and customized user reviews based on their behaviors on Twitter. For the text preprocessing based on keyword based approach the emotions and sentiments from the twitter data were used. Shivhare et. al [46] developed emotion detector system based on the emotion ontology produced an accuracy more than 75%.

Rahman et.al [47] proposed a methodology for sentence level emotion detection and created 25 emotion classes. It is based on the keyword analysis, emoticons, keyword negation, short word, a set of proverbs etc and achieved an accuracy of 80%.

4.2 Corpus based approach:

Corpus-based emotion detection approaches use supervised learning to induce sources of information such as word-emotion lexicons classified or weakly-labeled from a text corpus with a predefined collection of emotions extracted from emotion theories such as Ekman, Parrot, etc. To model the syntactic and semantic trends of text for emotion detection website such as wikipedia is used and unsupervised learning is also implemented. More works are focused on lexicons, motivated by a considerable amount of study in the area of sentiment analysis.

Anil et al[48] demonstrate how using a generative unigram mixture model (UMM) to jointly model emotionality and neutrality of terms, labelled (blogs, news headlines) and weakly labelled (tweets) emotion text can be used to learn a word-emotion interaction lexicon. UMM generated emotion language models (topics) have significantly lower perplexity compared to those from state-of-the-art generative models such as supervised Latent Dirichlet Allocation (sLDA).

Flor et al. [49] used multilingual dataset from tweets (English and Spanish). It consists of 8409 Spanish and 7303 English labeled dataset from the tweets. They reported linguistic statistics and applied machine learning to detect emotions. To evaluate the approaches used 10 fold validation and obtained accuracy of 64% for Spanish and 55%. For English. Rachman et al.[50] developed CBE (Corpus Based Emotions) with widely used emotion corpus the Wordnet Affect Emotions (WNA) and the Affective Norms for English Word (ANEW). They showed using the CBE, it improves the performance of detecting emotion with F-Measure using WNA and ANEW obtained is 50% and with the CBE obtained 61%.

Anil et al. [51] proposed the Unigram Mixture Model (UMM) based on the domain specific emotion lexicon. It outperforms the feature derived from supervised Latent Dirichlet (LDA) and Pointwise Mutual information (PMI) and also the combination of n-gram, lexion and POS. The F score measure for n-gram, PMI+UMM for total emotion intensity and for hybrid approach for the dataset such as SemEval, ISEAR, 10 and 5 cross validation is 38.23%, 39.48%, 6.24% and 52.18% respectively.

A constraint optimization architecture for emotion detection was developed by Yichen Wang and Aditya Pal[52]. To solve the multiple label classification and allow multiple emotions, their proposed model can be used. Their model achieved a precision of 0.43 for the Twitter dataset and a recall of 0.67.

4.3 Rule based Approach

To manipulate knowledge in order to view information in an advantageous way, the rule-based approach is used. It begins with text preprocessing initially, including stop word elimination, POS tagging, tokenization, etc. The rules of emotion are then derived using the concepts of statistics, linguistics, and computation. The best rules are selected later. Finally, the rules are applied to emotion datasets to determine the emotion labels. Subsequently, the appropriate rules are chosen. Additionally, the rules are applied to the emotion dataset for determining the emotion labels.

An alternative approach to improving the sentiment classification of user reviews in online communities is proposed by Asghar, Muhammad Zubair, et al.[53] Lexicon-enhanced sentiment analysis based on a rule-based classification and integrating emoticons, modifiers and domain-specific terms to evaluate the feedback posted in online reviews, in addition to the sentiment terms used in general purpose.

Dibyendu et.al [54] proposed sentence level emotion detection technique by applying semantic rules. It also includes negation words and obtained F1 score of 66.18%. Srinivas Badugu and Matla Suhasini [55] developed a Rule Based Approach that detects the emotions from the tweets and classify into different emotion categories and achieved an accuracy of 85%.

4.4 Machine learning approach

Emotion detection from text is based on classification problem involving different models from the disciplines of Natural Language Processing (NLP), Machine Learning(ML). Machine learning is categorized into unsupervised learning and supervised learning. Naive Bayes (NB), Support vector machine (SVM), conditional random field etc., are the most common traditional unsupervised machine learning.

Hasan et al. [56] in his work of detection of emotion worked on text stream data and use both online and offline messages. Support Vector Machine, Naïve Bayes and Decision Tree (DT) were used to detect emotions and achieved 90% accuracy. For YouTube comments, Tripto and Ali[34] suggested work on machine learning models and obtained classification of 59.2% accuracy and 65.97 % and 54.24% accuracy for Multiclass sentiment labels.

Merav et al.[57] proposed a model for the children with communication problem and help to find how to react to the societal situation. They used dataset which consist of non- insulting sentences (1241) and insulting sentences(1255). Applied ML algorithm and obtained 80% recall and more than 75 % precision for SVM method. And achieved precision and recall is (> 75%) for the Tree bagger, Multilayer Neural Network method.

Suhasini and Badugu[58] as implemented the machine learning approaches for the Twitter messages for the detection emotions. They showed efficiency in Naïve Bayes algorithm was more compared to the K Nearest Neighbour (KNN) and obtained an accuracy of 72.60%, 55.50% respectively. Fakhri[59] developed emotion recognition and prediction system for the detection of emotions using text. They used supervised learning algorithm such as Multinominal NB, DT, SVM and KNN for the ISEAR dataset and obtained highest accuracy for the Multinominal Naïve Bayes is 64.08%.

R Jayakrishna et al. [60] used machine learning approach for the Malyalam novel and using the SVM, classified sentences in the novel into happy and obtained precision of 0.94, for sad
obtained precision of 0.92, for fear obtained precision of 0.93, for anger obtained precision of 0.90 and lastly for surprise obtained precision of 0.90. Sonia and Kavitha [61] proposed an algorithm to identify the intensity of emotions from the Twitter dataset and identified the intensity of four types of emotions, considering tweets: {*happy, sad, angry and terror*}.

Wikarsa and Thahir [62] implemented machine learning Naïve Bayes algorithm for 105 tweets dataset and applied 10 cross validation to the approach and obtained accuracy of 83%. Forugh and Hooman,,[63] in their work vector similarity measure(VSM), keyword base and STASIS approaches are employed to detect emotions in text to find the different categories of emotions and obtained precision of 0.53.

4.4.1 Deep learning approach
Deep learning(DL) is a subset of ML in which programs learn from understanding and experiencing the hierarchy of concepts where each concept is described in terms of its relation to simpler concepts. This methodology helps a program to learn complex ideas by building them on simpler concepts [64].

In many research papers address DL model as long short-term memory (LSTM). LSTM consist of special type of RNN (recurrent neural network) with long-term dependency management capabilities. LSTM overcomes the issue of disappearing or bursting gradient prevalent in RNNs.

Baziotis et al.[65] propose work on SemEval 2018 Task 1 competition using Deep Learning model. Their idea was a two-layer Bidirectional long-term short-term memory (Bi-LSTM) built with a multilayer self-attention system. They used the approach with the tool ekphrasis[66] for pre-processing the text. Because of the limited number of training results, they used a transfer learning approach by pre-training the Bi-LSTMs on the SemEval-2017, Task 4A[67] dataset. A dataset of 550 million English tweets has been collected to be used in text preprocessing, word2vec embedding training [68],

IOP Conf. Series: Materials Science and Engineering 1110 (2021) 012009 doi:10.1088/1757-899X/1110/1/012009

and affective word embedding for calculating the required word statistics. The experimental results revealed that transfer learning did not outperform the random initialization model.

Zishan Ahmad et al. [69] designed a model using the DL classifier for Hindi text emotion detection and also showed that information gathered from resource-rich languages can be extended to other language domains using transfer learning and cross-lingual embedding.They obtained F1 score of 0.53.

Seo-Hui Park et.al [70] developed an emotion detection model using CNN and 144,701 tweets were used and also used a ROC story data. The Joy emotion was found with highest accuracy of 73.3% and Anger results in the lowest accuracy value of 36.7% and the lowest Kappa score is 0.216.

Xiao Zhang et al [71] introduced the Factor graph model is used to detect multiple emotions or Online Social Network and also proposed multilabel learning algorithm and achieved contextual information and obtained F1 score of 62.7, other categories F1 score for BR is 57.0, F1 score for Back Propagation Neural network is 57.7, F1 score for Probabilistic classifier chain is 54.1, F1 score for Label combination 55.7 and finally F1 score for Machine Learning K Nearest Neighbor is 56.0.

Malte and Ratadiya[72]developed BERT (Bidirectional Transformer). Both for Hindi and English text were tested and attained F1 score of 0.4521 and 0.5520 respectively. Waleed Ragheb et.al [73] developed a model using the SemEval2019 task 3 dataset.

The proposed model use deep transfer learning, self-attention mechanism and turn-based conversational modeling to classify the emotions and 0.7582 F1 score is obtained. Ma et al, [74] used Bi-LSTM to distinguish emotions into good, sad and angry in text and Emoji assertions and noticed that their method's performance exceeded the baseline models for happy and angry, but not for sad. Bi-LSTMs from documents are capable of extracting contextual knowledge. They obtained Micro F1 score is 0.7557.

Huang et al. [75] proposed Hierarchical LSTMs for the Contextual Emotion Detection (HRLCE) model and with BERT for detection of emotion using SemEval-2019 Task 3obtained the harmonic means score of 0.779 and ranked 5th among 165 Team. Daniel Haryadi, Gede Putra Kusuma [76] Classified seven emotions (anger, fear, joy, love, sadness, surprise, and thankfulness) using LSTM, nested LSTM. And Nested LSTM obtained the best accuracy of 99.167%, LSTM obtained 99.22% average precision, 98.86% of average recall and 99.04% of F1-score.

Mounika Karnaa et al. [77] used LSTM, SVM, NESTED LSTM. Average F1 rank obtained for LSTM is 94.1% and for Nested LSTM 92.2%. And LSTM obtained best average performance. And obtained F1score of 94.1% . Ankush et al. [22] participated in SemEval 2019 in that a training data set consists of 30160 dialogues, and two evaluation data sets were given as Test1 contains 2755 and Test 2 contains 5509 dialogues were given to the participants. Used Bidirection and Test2, which n LSTM approach and achieved F1 score is 79.59 micro. The best performances obtained for the class "Sad" and for the worst performance obtained for "Happy" emotion.

4.5 Hybrid Approach

In a unified model, the hybrid approach is a combination of different approaches. This approach has a higher likelihood of transcending the other approaches individually, leveraging the strength of the approaches used while trying to conceal their corresponding limitations.

Riahi and safari [79] proposed an approach for emotion detection in implicit texts. Based on three subsystems, they implemented a combinational framework. Every one analyses input information from a different perspective and produces as output an emotion label. A machine learning algorithm is the first subsystem. The second is a vector space model (VSM) based mathematical method, and the third is keyword based submodel with an information fusion component to aggregate the final output of main system. Their conclusions are aggregated and used to annotate the test text if it is otherwise left abandoned and only if all three subsystems agree on the same emotion type. The efficiency of the method proposed is 9.13 % higher than the machine learning subsystem, 16.6 percent better than VSM, and 23% better than the keyword-based method.

Ramalingam et al. [80] developed a hybrid model combined with the keyword and learning based method and obtained high accuracy result for detection of emotions from the text. Angelina et al [81] used twitter dataset with NRC emotion lexicon. And used SVM for multiclass classification and implemented on the software WEKA obtained accuracy of 84.92%, Spark obtained accuracy of 88.01%.

Hamed Khanpour, Cornelia Caragea [82] proposed system explains with or without lexicon system to develop health domain is often expensive. It also combines with CNN and LSTM model to capture the Hidden semantics in online health model. And also used ConvexLex LSTM and obtain high performance. There dataset contain 1066 sentences Cancer Survivors' Network (CSN) so it is represented as B-DS. And another set on the lung cancer discussion contains 1041 sentences and represent as L-DS found Joy and Sad sentence more compare to other emotions. And the highest F1 score of ConvLex LSTM for joy is 93.2 and 89.8 and for sad 92.3 and 89.4

Perikos and Hatzilygeroudis [83] developed model which classify emotions and found the performance was satisfactory. Used Naïve Bayes, maximum entropy, knowledge based tool, and Ensemble classifier used and obtained an accuracy of 77%, 85%, 80%, and 87% respectively

5. Evaluation Metrics:

Evaluation metrics are used to measure the statistics between the good models that can be fit. The most common metrics used to measure the models are Kappa Coefficient, multi label accuracy (Jaccard accuracy), F-Score, Precision and Recall, Accuracy, Pearson Correlation, 10 fold cross validation, Chi Square.

5.1 Kappa Coefficient: [70]

It is a statistical measure of an inter-annotator reliability or agreement. Kappa coefficient is used to assess qualitative documents and determine the agreement between two annotators. The equation (1) used to calculate kappa is:

$$K = \frac{po - pe}{1 - pe} = 1 - \frac{1 - po}{1 - pe} \qquad \text{Eq. (1)}$$

Where p0 is an annotator's relative observable agreement and pe is the hypothetical probability of chance agreement. Using the observed data, p0 and pe are calculated to determine the probability of each observer randomly informing each group. It ranges from 0 to 1.

5.2 Jaccard Accuracy: [86]

It defined as the size of the intersection divided by the size of the union of two label sets and is used to compare set of predicted labels for a sample to the corresponding set of labels in original. It ranges from 0 to 1. The following equation (2) is used to calculate Jaccard Accuracy is:

$$J(A,B) = \frac{|A \cap B|}{|A \cup B|} = \frac{|A \cap B|}{|A| + |B| - |A \cap B|} \qquad \text{Eq. (2)}$$

5.3 Precision, Recall, F-Score, Accuracy:

The Precision (P) defined as the number of true positives (T_p) over the number of true positives plus the number of false positives (Fp). [52], [57], [60], [63] .The following equation (3) is used to calculate the Precision:

$$P = \frac{Tp}{Tp + Fp} \qquad \text{Eq. (3)}$$

The Recall is defines as the number of true positives over the number of true positives plus the number of false negatives [52],[57]. The following equation (4) is used to calculate Recall is:

$$R = \frac{Tp}{Tp + Fn} \qquad \text{Eq. (4)}$$

where R represents Recall, Tp is True positive and Fn is False negative.

F-Score measure is used to provide a score that balances both the concerns of precision and recall in one number. And macro F1 score is used to measure when multiple classes are declared. MacroF1 score has best value =1 and worst value as 0. [22], [50], [51],

[71], [72], [73], [77], [69], [76], [82]. The following equation (5) is used to calculate the F1 score is:

$$F1 = 2*\frac{P \times R}{P+R} \quad \text{Eq. (5)}$$

Where P denotes as Precision and R denoted as Recall.

Accuracy metric is to measure the classification models. [46], [47], [49], [55], [56], [59], [61], [62], [81], and is calculated using the equation (6, 7):

$$\text{Accuracy}=\frac{\textit{Number of correct predictions}}{\textit{Total number of predictions}} \quad \text{Eq. (6)}$$

$$\text{Accuracy}=\frac{TP+TN}{TP+TN+FP+FN} \quad \text{Eq. (7)}$$

Where TP represent True positive, TN represents True negative, FP represents False positive, FN represents False Negative

5.4 *Pearson Correlation* [85]

It is the statistics that measure the statistical relationship or is the best method for an association, between the two variables that are continuous because it is based on the covariance of the two variables, and then it is divided by the product of their standard deviations. The following equation (8) is used to calculate Pearson Correlation is:

$$r = \frac{\sum(x_i - \overline{x})(y_i - \overline{y})}{\sqrt{\sum(x_i - \overline{x})^2 (y_i - \overline{y})^2}} \quad \text{Eq. 8}$$

Where r is a correlation coefficient, x_i is values of the x-variable in a sample, $\overline{x}$ is mean of the values of the x-variable, y_i is values of the y-variable in a sample and $\overline{y}$ is mean of the values of the y-variable

5.5 *10-fold cross validation:* [49],[51],[62]

The cross-validation technique is used to partitioning the original sample into a training set to train the model and a test set to validate it and to evaluate predictive models in machine learning. This procedure is named as k fold cross validation where the original data splits into k subsample. If a specific value is chosen for k, it can be used in the model reference instead of k, such as k=10 becomes a 10-fold cross-validation.

5.6 *Chi Square* [84]

To test the independence of two cases, a chi-square test is used in statistics. We can get observed count O and predicted count E given the data of two variables. Chi-Square tests how the predicted number E and the measured number O deviate from each other. The following equation (9) is used to calculate ChiSquare is:

$$x_c^2 = \sum \frac{(O_i - E_i)^2}{E_i} \quad \text{Eq. (9)}$$

Where c = degree of freedom, Observed value(s) and E=expected value(s)

6 Summary of existing approach

Table 4 discuss about the summary of the existing approaches, contribution and limitations.

Table 4: Summary of the recent existing approaches in text-based emotion detection

Proposed Work (Author, Year)	**Approaches Used**	**Contribution**	**Limitations**
Suhasini and Badugu, 2020 [58]	Machine Learning(ML)	Demonstrated the efficiency of the Naive Bayes(NB) in correlation with the K Nearest Neighbor (KNN). The accuracy obtained for NB is 72.06% and KNN 55.50%.	Low extraction of relevant data in sentences.
Ahmad Fakhri Ab. Nasir et.al 2020 [59]	Machine Learning(ML)	Comparisons of Four machine learning algorithm and developed graphical UI	Worked for Basic Model. Mixed emotions are not included
Seo-Hui Park et.al 2020 [70]	Deep Learning(DL)	Created an embedding emotion model using CNN, highest accuracy for joy of 73.3%, lowest accuracy value of 36.7% and the lowest Kappa score of 0.216	Negate sentence were not achieved.
Dibyendu et al, 2020 [54]	Rule Based	Used ISEAR dataset with specific regard to phrasal verbs to identify emotions.	Lack of contextual meaning. Insufficient words in vocabulary for the lexicon.
Mounika Karnaa et al. 2020 [77]	Deep Learning(DL)	Using LSTM, SVM, Nested LSTM identified Multiclass labels for emotion and achieved the accuracy F1=94.1, 92.9, 92.2 respectively	Recommended to use for complex dataset.
Flor Miriam Plaza-del-Arco et al. 2020 [49]	Corpus Based	Used Multilingual dataset and applied machine learning approach for detection of emotions from dataset and obtained accuracy for Spanish 0.64 and accuracy for English 0.55	Anger, disgust, fear, sad emotions are not able to be identified by the proposed system.
Xiao Zhang et.al 2020 [71]	Deep Learning(DL)	Factor graph model is used to detect multiple emotions or Online Social Network and also proposed multilabel learning algorithm and achieved contextual information. F1 score of 62.7	Limited annotated text is been used.

Daniel Haryadi, Gede Putra Kusuma 2020 [76]	Deep Learning(DL)	Classified emotions into seven they are: (fear, anger, love, joy, surprise, thankfulness and sadness) using LSTM AND nested LSTM. Obtained best accuracy for Nested LSTM is 99.167% and average precision and recall for LSTM is 99.22%, 98.86%, respectively and Obtained F1-score 99.04%.	Recommended to use more sophisticated Deep learning for the challenging dataset.
Hasan et al, 2019 [56]	Hybrid	Developed Emotex for online stream data and obtained accuracy of 90%	Semantic featured extraction should be improved.
Malte and Ratadiya 2019 [72]	Deep Learning(DL)	Constructed BERT (Bidirectional Transformer). Both for Hindi and English text were tested and attained F1 score for Hindi is 0.4521 and for English is 0.5520.	Hindered in performance due to presence of slang in Non English.
Waleed et al., 2019 [73]	Deep Learning (DL)	SemEval2019 dataset were used. The proposed model use deep transfer learning, self attention mechanism and turn-based conversational modeling to classify the emotions. Obtained F1 of 0.7582.	Failed to find the emotion –"happy"
Luyao et al, 2019 [74]	Deep Learning (DL)	Bidirectional LSTM was used to distinguish emotion oriented text into good, sad and angry and emoji's. Showed that the output of their method surpassed the baseline models {happy, angry} but failed for the sad. Micro F1 score of 0.7557	Limited types or classes of emotions
Ankush et al, 2019 [22]	Deep Learning(DL)	Developed Bidirectional LSTM and obtained a highest micro F score of 79.59	Not suitable for generalisation due to number of categories are finite.
Zishan Ahmad et al, 2019 [69]	Deep Learning (DL)	The information collected from resource-rich languages can be extended to other language domains by transfer learning and from the cross-lingual embedding. Obtained F1 of 0.53	Contextual meanings of word are disregarded.
Huang et al, 2019 [75]	Deep learning	Proposed Hierarchical LSTMs for the Contextual Emotion	Misclassification were high

		Detection (HRLCE) model and with BERT obtained the harmonic mean score of 0.779	
Forugh and Hooman, 2019 [63]	Machine Learning(ML)	Vector similarity measure (VSM), keyword base and STASIS approaches are employed to detect emotions in text to find the different categories of emotions and obtained precision of 0.53.	Weak qualitative extraction of information
Angelina et al,2019 [81]	Hybrid	To derive actionable emotion patterns in Tweets, the NRC emotion lexicon and SVM were used against WEKA, Spark software 84.92%, 88.01% accuracy were obtained.	Not suitable for generalisation due to the finite number of classes of emotions
R Jayakrishnan et al, 2018 [60]	Machine Learning(ML)	Using the SVM, classified sentences in the novel into happy and obtained precision of 0.94, for sad obtained precision of 0.92, for fear obtained precision of 0.93, for anger obtained precision of 0.90 and lastly for surprise obtained precision of 0.90	Disintegration of semantic information in texts document and Fail to understand the context of a word in the sentence.
Hamed Khanpour, Cornelia Caragea 2018 [82]	Hybrid	The proposed system explains with or without lexicon system to develop health domain is often expensive. It also combines with CNN and LSTM model to capture the Hidden semantics in online health model.F1 score of ConvLex LSTM (B-DS and L-DS cancer data)for joy 93.2 and 89.8 and for sad 92.3 and 89.4	Limited emotions are identified
Ramalingam et al.2018 [80]	Hybrid	Developed model combined with keyword-based and learning-based and obtained satisfactory results	Better classification techniques should be applied to improve performance.
Merav et.al 2018 [57]	Machine Learning[ML]	Developed automated model for communication problem with children and others for the better society assistance and to develop the mode applied machine learning algorithm and	Lack of semantic information

		SVM method: 80% recall and precision obtained more than 75%, and t Multi-Layer Neural Network and the Tree Bagger :precision and recall obtained more than 75%	
Matla, Suhasini, 2017 [55]	Rule Based	Four emotions with 85% accuracy were identified by the process.	Because of limited categories, the proposed methodology has stronger generalisation potential and Low contextual information extraction.
Sonia and Kavitha, 2017 [61]	Machine Learning(ML)	The intensity of four categories of emotions: happy, sad, angry and fear were identified.	A comparatively limited number of data. Limited number of categories of emotions. Disregarded for adjective strength
Anil Bandhakavia et. al 2017 [51]	Lexicon Based	Developed UMM Model based on the DSEL and outperforms the PMI and sLDA methods.	Only high quality lexicon can capture the context.
Rahman et al. 2017 [47]	Keyword based	Developed the method which solves the problem in sentence level emotion detection and in emoticons and achieved 80% of accuracy with 25 emotion classes, keyword analysis, short form of words, keyword negation analysis, emoticons set of proverbs and so on	Need to work future for paragraph sentences.
Perikos and Hatzilygeroudis, 2016 [83]	Hybrid	The performance of classifying emotions was satisfactory. Used NB, maximum entropy, knowledge based tool, and Ensemble classifier used. Obtained Accuracy of 77%, 85%, 80%, 87% respectively	Better classification techniques can be applied to improve the execution process.
Fika Hastarita Rachman et al. 2016 [50]	Corpus Based	Developed Corpus Based Emotion (CBE) with Wordnet Affect Emotion (WNA) and ANEW (Affective Norms for English Words improved performance with F score of 0.61	The label was seen only with highest frequency scattered.
Wikarsa and Thahir, 2015 [62]	Machine Learning[ML]	Five categories with 83 percent precision obtained	A better ML algorithm can enhance performance. And also considered small number of Tweets.

Shiv Naresh Shivhare et.al 2015 [46]	Keyword based	Keyword based technique is used and ontology approach helped to overcome semantic features. Proposed system achieved accuracy more than 75%	Determining Emotion Indicators is difficult
Yichen Wang,Aditya Pal 2015 [52]	Lexicon based	Developed optimized framework suitable for large dataset and multi label emotions are also identified. F measure of 0.63 for SemEval and ISEAR of 0.74	Proposed model misclassified for microblogs dataset

7 Conclusion

In this paper, a comparison of various approaches for detecting an individual's emotional state from textual data has been undertaken. The three major approaches of the emotion modeling in the psychology research such as Categorical approach, Dimensional approach and Appraisal based approaches was discussed. Further, different computational approaches proposed for emotion detection from text such as Keyword based approach, Rule based approach, Machine learning-based approaches and Hybrid approaches was discussed. It further explores existing state-of-the-art with focus on their approaches applied, evaluation measures, datasets used, signification contributions and limitations useful for budding researchers.

References

[1] Lutz, Catherine A. Unnatural emotions: Everyday sentiments on a Micronesian atoll and their challenge to Western theory. University of Chicago Press, 2011

[2] Gross, James J., et al. "Emotion and aging: Experience, expression, and control." *Psychology and aging* 12.4 (1997): 590.

[3] Kopelman, Shirli, Ashleigh Shelby Rosette, and Leigh Thompson. "The three faces of Eve: Strategic displays of positive, negative, and neutral emotions in negotiations." *Organizational Behavior and Human Decision Processes* 99.1 (2006): 81-101.

[4] Ekman, Paul Ed, and Richard J. Davidson. *The nature of emotion: Fundamental questions*. Oxford University Press, 1994.

[5] Garcia, David, and Frank Schweitzer. "Emotions in product reviews--empirics and models." 2011 IEEE Third International Conference on Privacy, Security, Risk and Trust and 2011 IEEE Third International Conference on Social Computing. IEEE, 2011.

[6] Barrett, L. F. 2006. "Solving the Emotion Paradox: Categorization and the Experience of Emotion," Personality and Social Psychology Review, vol. 10, pp. 20-46, 2006.

[7] Gunes H, Pantic M (2010) Automatic, dimensional and continuous emotion recognition. Int J Synth Emot (IJSE) 1(1):68–99.

[8] Ekman P (1999) Basic emotions. In: Dalgleish T, Power M (eds) The handbook of cognition and emotion. Wiley, New York, pp 45–60.

[9] Plutchik R. A general psychoevolutionary theory of emotion. Amsterdam, Netherlands: Elsevier; 1980 (pp. 3–33).

[10] Ortony A, Clore GL, Collins A. The Cognitive Structure of Emotions. Cambridge, MA: Cambridge University Press; 1990.

[11] Russell JA. A circumplex model of affect. J PersSoc Psychol. 1980;39(6):1161

[12] Russell JA, Mehrabian A. Evidence for a three-factor theory of emotions. J Res Pers. 1977;11(3):273-294.

[13] W. G. Parrott, Emotions in social psychology, Psychology Press, Philadelphia,2001.

[14] Kao, Edward Chao-Chun, et al. "Towards text-based emotion detection a survey and possible improvements." 2009 International Conference on Information Management and Engineering. IEEE, 2009.

[15] Aman S, Szpakowicz S (2007) Identifying expressions of emotion in text. In: Proceedings of the 10th international conference on text, speech and dialogue, TSD'07. Springer, Berlin, pp 196–205

[16] Alm CO, Roth D, Sproat R (2005) Emotions from text: machine learning for text-based emotion prediction. In: Proceedings of the conference on human language technology and empirical methods in natural language processing. Association for Computational Linguistics, Stroudsburg, PA, HLT '05, pp 579–586

[17] Scherer KR, Wallbott HG (1994) Evidence for universality and cultural variation of differential emotion response patterning. J Personal Soc Psychol 66(2):310–328

[18] Yadollahi, Ali, Ameneh Gholipour Shahraki, and Osmar R. Zaiane. "Current state of text sentiment analysis from opinion to emotion mining." *ACM Computing Surveys (CSUR)* 50.2 (2017): 1-33.

[19] Strapparava C, Mihalcea R (2007) Semeval-2007 task 14: affective text. In: Proceedings of the 4th international workshop on semantic evaluations, SemEval'07. Association for Computational Linguistics. Stroudsburg, pp 70–74

[20] Rosenthal S, Farra N, Nakov P. SemEval-2017 task 4: sentiment analysis in twitter; 2019. arXiv preprint arXiv:1912.00741

[21] Mohammad SM, Bravo-Marquez F, Salameh M, Kiritchenko S (2018) SemEval-2018 Task 1: affect in tweets. In: Proceedings of international workshop on semantic evaluation (SemEval-2018), New Orleans

[22] Chatterjee A, Narahari KN, Joshi M, Agrawal P. SemEval-2019 task 3: EmoContext contextual emotion detection in text. Paper presented at: Proceedings of the 13th International Workshop on Semantic Evaluation; 2019:39-48.

[23] Buechel S, Hahn U. Readers vs: writers vs. texts: coping with different perspectives of text understanding in emotion annotation. Paper presented at: Proceedings of the Proceedings of the 11th Linguistic Annotation Workshop; 2017:1-12.

[24] Klinger, Roman. "An analysis of annotated corpora for emotion classification in text." *Proceedings of the 27th International Conference on Computational Linguistics*. 2018

[25] Mohammad SM, Bravo-Marquez F. WASSA-2017 shared task on emotion intensity; 2017. arXiv preprint arXiv:1708.03700

[26] Alm ECO. Affect in* Text and Speech. Citeseer. Urbana: University of Illinois at Urbana-Champaign; 2008

[27] Li Y, Su H, Shen X, Li W, Cao Z, Niu S. Dailydialog: a manually labelled multi-turn dialogue dataset; 2017. arXiv preprint arXiv:1710.03957.

[28] Ghazi D, Inkpen D, Szpakowicz S. Detecting emotion stimuli in emotion-bearing sentences. Paper presented at: Proceedings of the International Conference on Intelligent Text Processing and Computational Linguistics; 2015:152-165; Springer.

[29] Poria S, Hazarika D, Majumder N, Naik G, Cambria E, Mihalcea R. MELD: a multimodal multi-party dataset for emotion recognition in conversations; 2018. arXiv preprint arXiv:1810.02508.

[30] Wang B, Liakata M, Zubiaga A, Procter R, Jensen E. Smile: Twitter emotion classification using domain adaptation. Paper presented at: Proceedings of the 25th International Joint Conference on Artificial Intelligence; 2016:15; AAAI.

[31] Liu C, Osama M, DeAndrade A. DENS: a dataset for multi-class emotion analysis; 2019. arXiv preprint arXiv:1910.11769

[32] Preo¸tiuc-Pietro D, Schwartz HA, Park G, et al. Modelling valence and arousal in facebook posts. Paper presented at: Proceedings of the 7th Workshop on Computational Approaches to Subjectivity, Sentiment and Social Media Analysis; 2016:9-15

[33] Liu V, Banea C, Mihalcea R. Grounded emotions. Paper presented at: Proceedings of the 2017 7th International Conference on Affective Computing and Intelligent Interaction; 2017:477-483; IEEE.

[34] Tripto, Nafis Irtiza, and Mohammed Eunus Ali. "Detecting multilabel sentiment and emotions from bangla youtube comments." *2018 International Conference on Bangla Speech and Language Processing (ICBSLP)*. IEEE, 2018

[35] Nielsen FÅ (2011) A new ANEW: evaluation of a word list for sentiment analysis in microblogs. In: Proceedings of the ESWC2011 workshop on 'Making Sense of Microposts': big things come in small packages. Heraklion, Crete, pp 93–98

[36] Kiritchenko S, Zhu X, Mohammad SM (2014) Sentiment analysis of short informal texts. J Artif Intell Res (JAIR) 50(1):723–762

[37] Mohammad, Saif M., et al. "Sentiment, emotion, purpose, and style in electoral tweets." *Information Processing & Management* 51.4 (2015): 480-499.

[38] Mohammad SM, Kiritchenko S (2015) Using hashtags to capture fine emotion categories from tweets. Comput Intell 31(2):301–326

[39] Hu M, Liu B (2004) Mining and summarizing customer reviews. In: Proceedings of the 10th ACM SIGKDD international conference on knowledge discovery and data mining, KDD'04. ACM, New York, pp 168–177

[40] Kiritchenko S, Zhu X, Mohammad SM (2014) Sentiment analysis of short informal texts. J Artif Intell Res (JAIR) 50(1):723–762.

[41] Fellbaum C (1998) WordNet: an electronic lexical database. Language, speech, and communication. MIT Press, Cambridge

[42] Mohammad SM, Turney PD (2010) Emotions evoked by common words and phrases: using mechanical turk to create an emotion lexicon. In: Proceedings of the NAACL HLT 2010 workshop on computational approaches to analysis and generation of emotion in text, CAAGET'10. Association for Computational Linguistics, Stroudsburg, pp 26–34.

[43] Mohammad SM, Turney PD (2013) Crowdsourcing a word-emotion association lexicon. Comput Intell 29(3):436–465.

[44] Shah, Faisal Muhammad, et al. "Emotion Detection from Tweets using AIT-2018 Dataset." *2019 5th International Conference on Advances in Electrical Engineering (ICAEE)*. IEEE, 2019.

[45] Sailunaz, Kashfia, and Reda Alhajj. "Emotion and sentiment analysis from Twitter text." Journal of Computational Science 36 (2019): 101003.

[46] Shivhare, Shiv Naresh, Shakun Garg, and Anitesh Mishra. "EmotionFinder: Detecting emotion from blogs and textual documents." *International Conference on Computing, Communication & Automation*. IEEE, 2015.

[47] Rahman, Romana. "Detecting emotion from text and emoticon." *London Journal of Research in Computer Science and Technology* (2017).

[48] Bandhakavi, Anil, et al. "Lexicon generation for emotion detection from text." *IEEE intelligent systems* 32.1 (2017): 102-108.

[49] Del Arco, Flor Miriam Plaza, et al. "EmoEvent: A multilingual emotion corpus based on different

events." *Proceedings of The 12th Language Resources and Evaluation Conference*. 2020.

[50] Rachman, Fika Hastarita, Riyanarto Sarno, and Chastine Fatichah. "CBE: Corpus-based of emotion for emotion detection in text document." *2016 3rd International Conference on Information Technology, Computer, and Electrical Engineering (ICITACEE)*. IEEE, 2016.

[51] Bandhakavi, Anil, et al. "Lexicon based feature extraction for emotion text classification." *Pattern recognition letters* 93 (2017): 133-142.

[52] Wang, Yichen, and Aditya Pal. "Detecting emotions in social media: A constrained optimization approach." *Twenty-fourth international joint conference on artificial intelligence*. 2015.

[53] Asghar, Muhammad Zubair, et al. "Lexicon-enhanced sentiment analysis framework using rule-based classification scheme." *PloS one* 12.2 (2017): e0171649.

[54] Seal D, Roy UK, Basak R. Sentence-level emotion detection from text based on semantic rules. Paper presented at: Proceedings of the Information and Communication Technology for Sustainable Development; 2020:423—430; Springer

[55] Badugu S, Suhasini M. Emotion detection on twitter data using knowledge base approach.International Journal of Computer Applications. 2017;162(10):28-33.

[56] Hasan M, Rundensteiner E, Agu E. Automatic emotion detection in text streams by analyzing twitter data. Int J Data Sci Anal. 2019;7(1):35-51

[57] Allouch, Merav, et al. "Automatic detection of insulting sentences in conversation." *2018 IEEE International Conference on the Science of Electrical Engineering in Israel (ICSEE)*. IEEE, 2018.

[58] Suhasini M, Srinivasu B. Emotion Detection Framework for Twitter Data Using Supervised Classifiers. New York, NY: Springer; 2020 (pp. 565–576)

[59] Nasir, Ahmad Fakhri Ab, et al. "Text-based emotion prediction system using machine learning approach." *IOP Conference Series: Materials Science and Engineering*. Vol. 769. No. 1. IOP Publishing, 2020.

[60] Jayakrishnan R, Gopal Greeshma N, Santhikrishna MS. Multi-class emotion detection and annotation in malayalam novels. Paper presented at: Proceedings of the 2018 International Conference on Computer Communication and Informatics; 2018:1-5; IEEE.

[61] Mashal SX, Asnani K. Emotion intensity detection for social media data. Paper presented at: Proceedings of the 2017 International Conference on Computing Methodologies and Communication; 2017:155-158; IEEE.

[62] Wikarsa L, Thahir SN. A text mining application of emotion classifications of Twitter's users using Naive Bayes method. Paper presented at: Proceedings of the 2015 1st International Conference on Wireless and Telematics (ICWT); 2015:1-6; IEEE.

[63] Mozafari F, Tahayori H. Emotion detection by using similarity techniques. Paper presented at: Proceedings of the 2019 7th Iranian Joint Congress on Fuzzy and Intelligent Systems; 2019:1-5; IEEE

[64] Goodfellow I, Bengio Y, Courville A (2016) Deep learning. MIT Press, Cambridge.

[65] Baziotis, Christos, et al. "Ntua-slp at semeval-2018 task 1: Predicting affective content in tweets with deep attentive rnns and transfer learning." *arXiv preprint arXiv:1804.06658* (2018).

[66] Baziotis C, Pelekis N, Doulkeridis C (2017) Datastories at semeval-2017 task 4: deep LSTM with attention for message-level and topic-based sentiment analysis. In: Proceedings of the 11th international workshop on semantic evaluation (SemEval-2017). Association for Computational Linguistics, pp 747– 754

[67] Rosenthal S, Farra N, Nakov P (2017) SemEval-2017 task 4: sentiment analysis in twitter. In: Proceedings of the 11th international workshop on semantic evaluation. Association for Computational Linguistics, Vancouver, SemEval'17.

[68] Mikolov T, Sutskever I, Chen K, Corrado G, Dean J (2013) Distributed representations of words

and phrases and their compositionality. In: Proceedings of the 26th international conference on neural information processing systems, Vol 2, NIPS'13. Curran Associates Inc., USA, pp 3111–3119

[69] Ahmad Z, Jindal R, Ekbal A, Bhattachharyya P. Borrow from rich cousin: transfer learning for emotion detection using cross lingual embedding. Expert Syst Appl. 2020;139:112851.

[70] Park, Seo-Hui, Byung-Chull Bae, and Yun-Gyung Cheong. "Emotion Recognition from Text Stories Using an Emotion Embedding Model." *2020 IEEE International Conference on Big Data and Smart Computing (BigComp)*. IEEE, 2020.

[71] Zhang, Xiao, et al. "Emotion Detection in Online Social Networks: A Multilabel Learning Approach." *IEEE Internet of Things Journal* 7.9 (2020): 8133-8143.

[72] Malte A, Ratadiya P. Multilingual cyber abuse detection using advanced transformer architecture. Paper presented at: Proceedings of the TENCON 2019-2019 IEEE Region 10 Conference; 2019:784-789; IEEE.

[73] Ragheb W, Azé J, Bringay S, Servajean M. Attention-based modeling for emotion detection and classification in textual conversations; 2019. arXiv preprint arXiv:1906.07020.

[74] Ma L, Zhang L, Ye W, Hu W. PKUSE at SemEval-2019 Task 3: emotion detection with emotion-oriented neural attention network. Paper presented at: Proceedings of the 13th International Workshop on Semantic Evaluation; 2019:287-291.

[75] Huang C, Trabelsi A, Zaïane OR. ANA at SemEval-2019 Task 3: contextual emotion detection in conversations through hierarchical LSTMs and BERT; 2019. arXiv preprint arXiv:1904.00132

[76] Haryadi, Daniel, and Gede Putra Kusuma. "Emotion Detection in Text Using Nested Long Short-Term Memory." *IJACSA) International Journal of Advanced Computer Science and Applications* 10.6 (2019).

[77] Karna, Mounika, D. Sujitha Juliet, and R. Catherine Joy. "Deep learning based Text Emotion Recognition for Chatbot applications." *2020 4th International Conference on Trends in Electronics and Informatics (ICOEI)(48184)*. IEEE, 2020.

[78] Ochsner, Kevin N., and James J. Gross. "The cognitive control of emotion." *Trends in cognitive sciences* 9.5 (2005): 242-249.

[79] Riahi N, Safari P (2016) Implicit emotion detection from text with information fusion. J Adv Comput Res 7(2):85–99

[80] Ramalingam, V. V., et al. "Emotion detection from text." *Journal of Physics: Conference Series*. Vol. 1000. No. 1. 2018.

[81] Tzacheva A, Ranganathan J, Mylavarapu SY. Actionable Pattern Discovery for Tweet Emotions. Paper presented at: Proceedings of the International Conference on Applied Human Factors and Ergonomics; 2019:46-57; Springer.

[82] Khanpour, Hamed, and Cornelia Caragea. "Fine-grained emotion detection in health-related online posts." *Proceedings of the 2018 Conference on Empirical Methods in Natural Language Processing*. 2018.

[83] Ochsner, Kevin N., and James J. Gross. "The cognitive control of emotion." *Trends in cognitive sciences* 9.5 (2005): 242-249.

[84] Perikos I, Hatzilygeroudis I. Recognizing emotions in text using ensemble of classifiers. Eng Appl Artif Intell. 2016;51:191-201

[85] Strapparava, Carlo, and Rada Mihalcea. "Semeval-2007 task 14: Affective text." *Proceedings of the Fourth International Workshop on Semantic Evaluations (SemEval-2007)*. 2007.

[86] Zhang, Xiao, et al. "Emotion Detection in Online Social Networks: A Multilabel Learning Approach." *IEEE Internet of Things Journal* 7.9 (2020): 8133-8143.

EmoCure: Utilising Social Media Data and Smartphones to Predict and Cure Depression

Ashish Basantani, Yash Kesarwani, Srishti Bhatia, and Sarika Jain
National Institute of Technology, Kurukshetra

abasantani_51810032@nitkkr.ac.in, yash_51810038@nitkkr.ac.in,
srishti_51810002@nitkkr.ac.in, jasarika@nitkkr.ac.in

Abstract. Depression is a common and recurring illness. It is a serious medical condition which might lead to self-harm. Various factors like unawareness about depression, shortage of medical health professionals and social stigmas make proper treatment for depression inaccessible. With the growing importance of technology in every sector of life, healthcare systems have also started using technology to provide better treatment. Various studies have shown that the widespread use of smartphones can be useful in predicting as well as treating depression by recommending activities. Smartphones can be very useful in the continuous monitoring of a patient which in turn helps to keep track of the activities of the patient. Social media data can also be used to find out the mental state of a patient. We propose EmoCure, a smartphone application that uses social media data, wearable sensor data, patient history, and smartphone usage patterns to predict, monitor, and treat depression using emotion regulating activities. We use machine learning models for finding out the sentiments in the social media posts. To predict depression, we use ensemble learning. We then recommend personalized emotion regulating activities whichever the user prefers.

Keywords: Activity Recommendation, Depression, Mobile Sensors, Prediction, Social Media Data

1. Introduction

Depression has become a worldwide problem with more than 300 million people suffering from it, as reported in a research done by World Health Organization (WHO) 0. According to a study done by WHO, 7.5 percent of the Indian population suffers from some or other form of mental disorder. In the year 2015, India recorded the highest number of cases of depressive disorders in the world 0. Severe depression for a continuous period can also lead to suicide. Various healthcare workers are working consistently to find out a proper treatment of depression. There are various social stigmas attached with depression that inhibit the proper treatment of depression. The authors in 0 prove that social stigma is the prominent reason that leads to depression being untreated. Unawareness about disease is

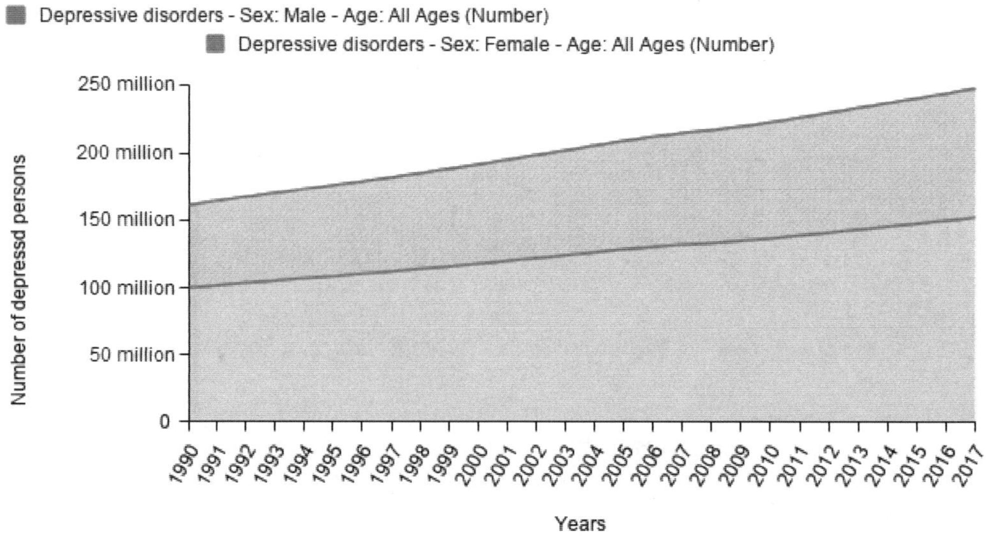

Figure 1.1. No. of people suffering from depression from 1990 to 2017

also one of the factors. The rise in the number of people suffering from depression from 1990 to 2017 can be seen from the below image (Fig 1.1).

Earlier, the healthcare workers asked a patient to be physically present at their clinic for receiving any type of recommendations about health. Mostly, the suggestions that doctor gave were based on the description given by the patient, symptoms that he can observe and the lab test results. The symptoms however can be misleading or not very well described, due to which the physicians may infer wrongly and may not be able to properly predict the patient's condition. The situation is even worse, where a patient has a mental disorder. There were no systems available earlier to keep track of the patient if he is not physically present at the doctor's clinic. Medical experts will be able to provide better treatment to the patients, if the patient can somehow be monitored remotely and continuously.

The rapid advancement in mobile phone technology and use of social media has presented an opportunity for providing easily accessible and affordable healthcare. Mobile phones are devices that monitor our activities 24x7. Mental health behavior can be assessed by continuous monitoring of the patient via sensors and via social media. Various studies have been done that explore the idea of using mobile phones and social media for providing patient-centric healthcare 0.

The authors in 0 present a framework to use mobile phones for delivering healthcare. The use of smartphones has also given rise to the use of social media. There are around 300 million users of Facebook 0, 33 million users of Twitter and 26 million users of LinkedIn 0. People spend a lot of their time on social media daily. On an average, people spend 2.5 hours a day on social media globally. Social media platform has become a medium to share activities, feelings, etc. A lot of people now believe in sharing feelings, making friends, sharing common interest and ideas on virtual medium more than the real life. Facebook, twitter, reddit, etc. has become an outlet for the emotions for people. Thus, analyzing social media posts can be very helpful in determining the mental state of a person.

Munmun De Choudhury et al. used this approach for understanding depression in individuals 0 and populations 0 and found social media to be advantageous in curing depression.
We have studied the already existing systems which aimed to predict and cure depression. We found out the various gaps that were present in the existing systems. We then propose our system that performs prediction of depression as well as monitoring and activity recommendation for helping patients suffering from depression. We discuss various machine learning algorithms that we used for prediction of depression and which resulted into the best one. We also discuss the recommendation methodology that we are implementing into our system.

The proposed system aims to provide a cost-effective cure for depressive disorders using smartphones and social media data by continuous monitoring and recommending emotion regulating activities. The objectives of this study are as follows:

- To predict depression using smartphone usage patterns, patient history, wearable sensor data, and social media data.
- To monitor the patients in daily life by collecting their smartphone usage patterns, wearable sensor data, and social media data.
- To provide patient-centric healthcare and cure depression through self-therapy by using data collected during monitoring to recommend emotion regulating activities.

2. Related Work

With the advent of technology, mobile phones have become an important part of our day to day life. From listening music, surfing Internet to navigating to a new location, everything is possible using a smartphone. Smartphones giving so much options makes us carry it wherever we go. These devices can hence become a medium to provide patient-centric healthcare. The authors in 0 introduce the concept of Augmented Personalized Healthcare (APH). According to APH, the prediction and treatment of depression requires continuous monitoring of a person and for this purpose a smartphone is the best option available.

A reason behind the extensive use of mobile phones is social media. Nowadays, people spend a lot of their time on social media websites such as Facebook, Twitter, Reddit and Instagram. People tend to share their thoughts and feelings on these virtual platforms more often than in real life. Analysis of a person's social media data can be quite effective input in predicting their mental state. By doing continuous monitoring of a person’s activities, we can determine the existence of the key symptoms of depression, like isolation, decreased activity, sleeping problems, etc. Hence to predict the onset of depression in a person, smartphone usage patterns and smartphone sensor data can be effectively used. Apart from prediction, continuous monitoring through mobile phones can prove to be very useful in recommending activities too.

MoodScope 0 developed by Microsoft, is a smartphone software to predict the mood of the user based on various inputs like text messages, emails, call logs, application usage, web browsing, and location. MoodScope is a mood logger application that is used by participants to self-report emotions. The data that is taken as an input is used to train a personalized mood model for each user using Multi-Linear regression. The personalized model achieved an accuracy of 93%, while an All-User mood model that can be used as a general model for all users initially achieved an accuracy of 66%. The study also develops a hybrid model, which achieves an accuracy of 72%. Galen Chin-Lun Hung et al. 0 build their system using a similar approach. They used mobile phone usage patterns to train four classifiers which helped to predict negative emotions. The authors found that the Naïve Bayes classifier with greedy best forward feature selection had the highest accuracy of 86.71%.

A research done by Munmun De Choudhary et al. 0 uses crowdsourcing to collect tweets of individuals from Twitter who reported that they have been diagnosed with clinical depression. The authors trained a Support Vector Machine (SVM) classifier using various attributes such as the number of tweets per day, the number of replies, the pattern of tweets, linguistic styles, etc. The SVM classifier achieved an accuracy of ~70% with 0.74 precision. Working on similar approach Johannes C. Eichstaedt et al. 0 used Facebook posts to predict depression in medical records. They trained logistic regression models to achieve a high accuracy of 72%. The study 0 aims to develop a web application that provides the user an easy-to-use medium to check their depression levels using their social media posts. The web application also provides a doctor's location nearest to user’s location, if the person is found to be suffering from depression. Facebook posts along with questionnaires filled by the user was given as an input to Naïve Bayes classifier to detect depression. The authors 0 predict depression using social media data. They captured publicly available Facebook data of bipolar, depression, and anxiety pages that contained users' comments using the NCapture tool. They train various classifiers and found that the Decision Tree classifier gives the best accuracy for their dataset. Gaur M et al. 0 used information (subreddits) from Reddit to check about suicidal tendencies and other mental health issues afflicting depressed users. The authors have built suicide risk severity lexicon using medical knowledge bases and suicide ontology to detect suicidal thoughts and actions. They found that for a 5-label confusion matrix CNN correctly classifies users with an accuracy of 80% and for a 4-label confusion matrix, it has 92% accuracy.

Galen Chin-Lun Hung et al. 0 developed an activity recommender system based on smartphone usage patterns, personal information, and environmental sensor information. The authors used clustering and combinational cube matrix (CCC) to recommend similar activities to similar users.

Intellicare is a suite of apps that aims to treat depression using interactive skill-based training that is based on cognitive-behavioral therapy, problem-solving therapy, etc. These apps were available on the Google Play Store. The study 0 evaluates the efficiency of Intellicare in reducing depressive symptoms. Shiqi Yang et al. 0 develop emHealth, an intelligent activity recommender system with depression prediction for emotion regulation. The authors first predict depression using decision tree technique and SVM as the prediction models and compare their accuracies. After detecting depression, the system recommends activities based on external factors of depression and the level of depression.

Fabian Wahle et al. 0 developed Mobile Sensing and Support (MOSS), a smartphone app that provides just-in-time interventions derived from cognitive behavior therapy. They use features such as the geographic movement, number of calls, time at home, number of calendar events, etc. to determine the actual behavior of the user and recommend activities accordingly.

The existing systems had a number of gaps. Firstly, the existing system use insufficient inputs to predict and cure depression. Most systems only use social media or smartphone usage patterns for prediction and activity recommendation. Using only these inputs leads to inaccurate results. Secondly, most systems use Twitter for monitoring the social activity of the user. Using Twitter data may lead to fetching irrelevant posts on the basis of certain keywords. Most systems that we studied didn't provide a single framework for both depression prediction and activity recommendations.

3. Methods

3.1. System Overview

The proposed system covers all the gaps in the existing systems. The system uses both social media data and smartphone usage patterns to predict and cure depression. The system also uses patient history and wearable sensor data. The system uses Reddit data instead of Twitter data to monitor social activity of the patients. Reddit is one of the most popular websites where people discuss topics, and there is a specific subreddit for each subject. It has a specific depression subreddit where most of the posts are from patients suffering from depression or mental health professionals. The system also provides a one-stop solution for predicting depression and activity recommendation.

The system comprises of three modules: Prediction, Monitoring and Recommendation. Figure 3.1 shows the system architecture. The Prediction module uses social media data, smartphone usage patterns, patient history and wearable sensor data predict whether the user is depressed or not. The Monitoring module is responsible for continuously monitoring the users' current mood. The Recommendation module is responsible for recommending mood regulating activities to the user. When the user logs in to the EmoCure application, a dashboard containing three options i.e., Prediction, Monitoring, and Recommendation are shown. Figure 3 shows the dashboard of the EmoCure application. The user can click on these options to perform the corresponding activity.

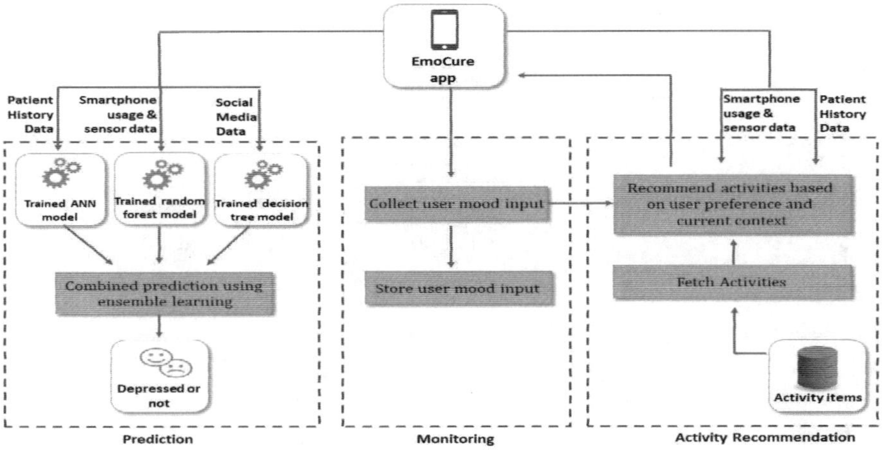

Figure 3.1. System architecture

The system performs prediction and activity recommendation using the following inputs:

- Social Media Data: The data that a person shares on their social media account can be used to figure out his state of mind. This data can be used effectively when the patient has a mental illness such as depression.
- Patient History: Patient's history will be beneficial in recommending better activities to the patients. For example, a person who was diagnosed with depression in the past has higher chances of being depressed again.
- Smartphone usage patterns: Smartphone usage patterns such as call logs, text messages can describe how much the user interacts with people. They provide insights into the daily activity of the users.
- Wearable Sensor Data: Wearable sensors such as Fitbit describe the physical activity of the users, i.e., the number of steps they have walked, how much exercise they have done, heart rate, etc. Smartphones are also equipped sensors such as GPS sensors that describe the places they have visited in a day.

3.2. Data Collection & Pre-processing

The inputs provided to the system are collected in the following ways:

Figure 3.2. EmoCure application dashboard

3.2.1 Social media data

For training the depression prediction model, we collected the social media data from Reddit. Reddit is one of the most popular websites where people discuss topics, and there is a specific subreddit for each subject. The popularity and the content length make it ideal for prediction. It has a specific depression subreddit where most of the posts are from patients suffering from depression or mental health professionals. We use the Reddit Pushshift API to extract Reddit submissions from the depression subreddit. The API is available on the Internet free of cost.

To pre-process the Reddit posts, we need to find the overall sentiment of each post. The sentiment of each post is calculated by first tokenizing the post into words. The sentiment of the post is the summation of the polarity of the tokenized words divided by the total number of words in the post. To find the polarity of the words, we downloaded a dictionary 0 which contains words of English language with their corresponding polarity i.e. negative or positive. For example, words like stress, sad and death are negative words whereas words like happy, excited and good are positive words. The overall sentiment of each post is calculated and stored with a unique ID assigned to the post in a CSV file. The dataset we collected contains 9331 Reddit posts.

3.2.2 Patient history data

For training the depression prediction model, we collected the patient history data from the Internet 0. The dataset contains attributes such as expenses, income, diseases, no of children, and other factors that can lead to depression for the farmers of South Africa.

The dataset initially had 76 attributes. We performed feature elimination manually by discarding attributes such as wage and livestock, which we found to be irrelevant. The dataset finally has 20 features. The dataset initially had 1143 instances. We discard those instances for which 50% of the attributes had missing or no values. The dataset finally has 1100 instances. The survey date is in the string format, so it is converted to a number using Label encoder.

The users' patient history data is collected using the EmoCure mobile phone application. The user enters attributes such as age, working hours, income, medical conditions etc. which represent the history and the lifestyle of the patient.

3.2.3 Smartphone usage patterns

Smartphone usage patterns such as app usage, location data, calls, etc. are required for activity recommendation based on current context. The EmoCure android application collects this data while monitoring the user. Permissions for collecting and storing each type of data are taken from the user.

3.3. Prediction

For prediction that a person is suffering from depression or not requires the complete knowledge of the person, from his social background in the real world to his social presence in the virtual world. Various research have shown that the social media can be very useful tool in determining whether a person is suffering from depression or not. To track his real-world activities continuous monitoring of people can also be done using the smartphones. The key symptoms of depression, such as isolation, decreased activity, and sleeping problems can be detected by continuous monitoring of a person's activities through wearable sensors and mobile phones.

We use ensemble learning to predict depression. Using ensemble learning, we can train different classifiers for the inputs and combine their outputs to give the final prediction. This enables us to use classifiers that are best suited for individual inputs. Figure 3.3 represents the depression prediction module.

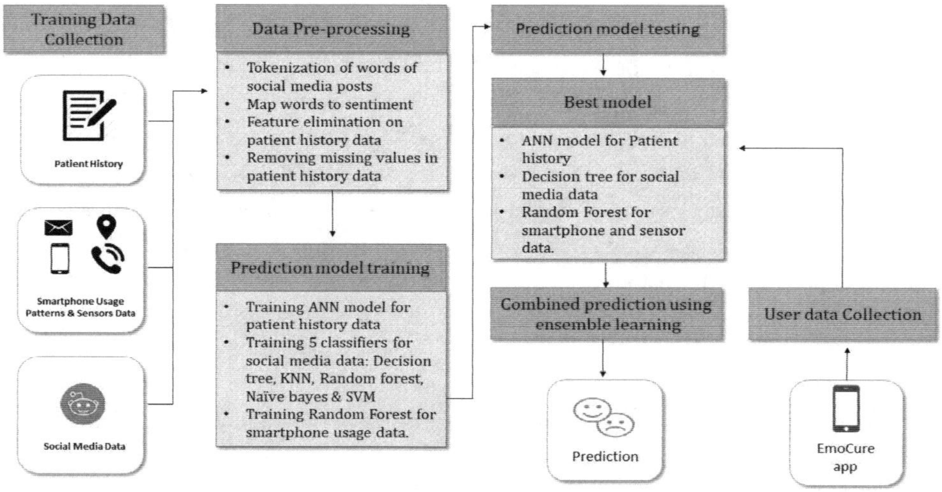

Figure 3.3. Prediction module

3.3.1 Social Media Data

We use machine learning methods to train the depression prediction model using social media data. We fetched the Reddit posts and their corresponding sentiment using the unique ID of each post for training the depression prediction model using social media data. We trained models using five different machine learning techniques, namely Decision Tree, Random Forest, Naïve Bayes, k-Nearest Neighbour, and Support Vector Machine. We found that of all the five algorithms, the decision tree gives the best accuracy. The Naïve Bayes method gave an accuracy of 93.79%, Support Vector Machine gave 50.0%, K-Nearest Neighbour gave 81.46%, Random Forest gave 49.1%, and Decision Tree gave the best accuracy of 98.55%. The sentiment analysis process is shown in Figure 4.4.

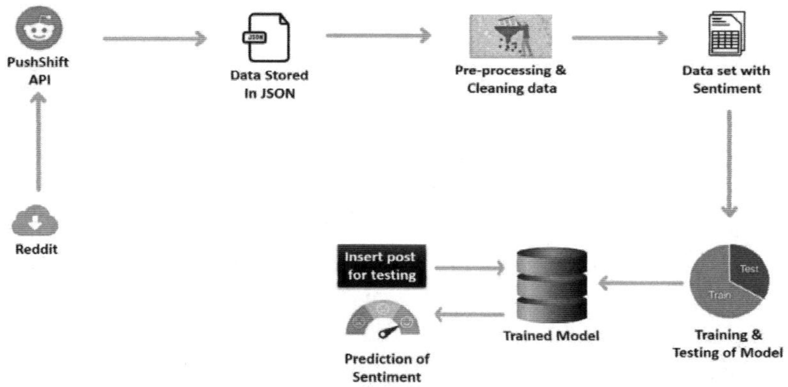

Figure 3.4. Social data analysis

3.3.2 Patient History Data

For training the classifier for patient history data, we use neural networks. We use neural networks because traditional machine learning algorithms might fail in the future when the dataset grows in size, and neural networks can efficiently handle large data. The neural network for prediction has two hidden layers, with thirty-six neurons in each layer. We use 70% of the rows in the dataset for training and 30% for testing. The train_test_split python function does this for us. With 100 epochs, the model achieved an accuracy of 86.7%.

3.4. Monitoring

Monitoring the users' mood is essential for tracking and controlling symptoms of depression in the user. It helps to understand the actual situation of the user. The MoodCheck module in the EmoCure application is responsible for monitoring the user. The MoodCheck module sends a notification to the user every four hours and asks her to input her current mood. The user can select a mood rating from 1 to 10 as shown in Figure 4.5. Mood information for every user is stored in the database. The MoodCheck module also shows the mood graph of the user for the day. The mood graph helps the users to track their own progress and their moods over the day. The user can also view their mood graph of any previous dates.

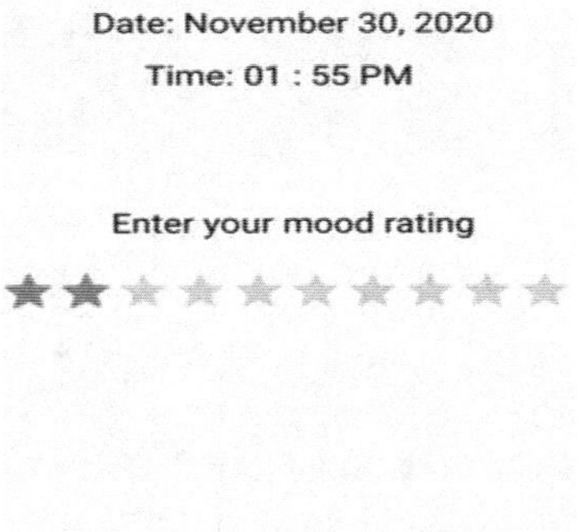

Figure 3.5. Enter user mood rating

Figure 3.5 shows the mood graph. Based on the current context and the current mood of the user, the application recommends mood regulating activities to the user.

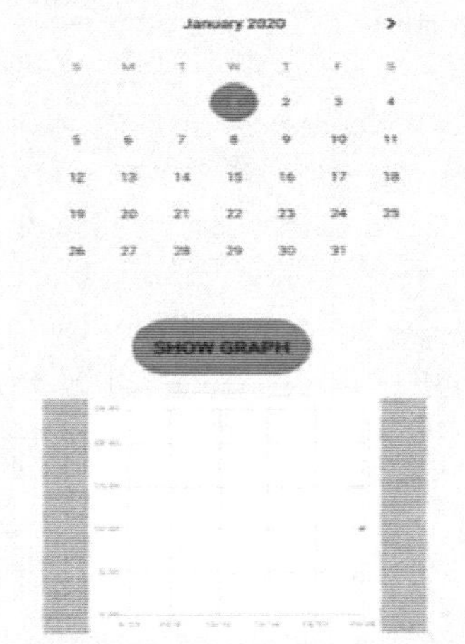

Figure 3.6. Mood graph

3.5. Activity Recommendation

The system performs activity recommendations based on the user's interaction with the EmoCure application. The activity recommendation module takes into account the current context, user's current mood and user's past preferences. Initially, the activities are recommended randomly from the set of activities that psychiatrists recommend to their patients suffering from depression. After days of monitoring and collecting user input, the system recommends the activities which were preferred by the user in the past. Figure 4.7 represents the activity recommendation module.

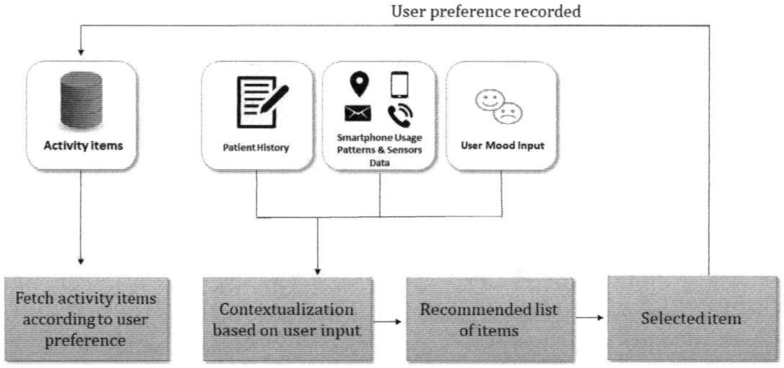

Figure 3.7. Recommendation module

Whenever the user gives a negative mood input to the MoodCheck module, the app recommends mood regulating activities based on the current context and user's past preferences. When the user gives a positive mood input, the application shows positive messages and recommends light activities such as listening to happy music.

For storing user preference data, a table is maintained for the activities in the database. The rows of the table depict the activities that can be recommended, and the columns depict one-hour time slots in a day. Each entry in the table corresponds to the number of times the user has chosen the particular activity during that particular time slot. Every time the user selects an activity the corresponding entry in the table increases. At the time of recommending an activity, the application will first fetch activities based on the maximum value in the table for the current time. The application will then perform contextualization by checking for multiple factors that affect the recommendation. Factors like environmental factors such as current weather, personal factors such as time constraints and working hours etc. represent the current situation of the user. These factors determine what activity can be recommended. For example, outdoor activities cannot be recommended when the current weather is rainy or snowy. Table 1 represents some of the activities and the factors that determine whether that activity can be recommended or not.

Table 1. List of activities

S.No.	Activity	Description	Conditions
1.	Social support	Suggest to connect with favorite contacts.	• Calendar is free for the current date and time.
2.	Self-affirmations	Suggest practicing self-affirmations, read positive quotes/articles or life lessons.	
3.	Exercise for mood	Suggest exercises.	• Non-working hours. • Calendar is free for the current date and time. • Weather conditions are favorable. • No disabilities/conditions that forbid exercise.
4.	Behavioral activation	Track positive streaks, mood history.	
5.	Mindfulness practice	Suggest different types of meditations.	• Non- working hours. • Calendar is free for the current date and time

After fetching highly preferred activities and performing contextualization based on user's current situation and lifestyle, the system will recommend three activities to the user. The user can choose any which he/she finds comfortable to do.

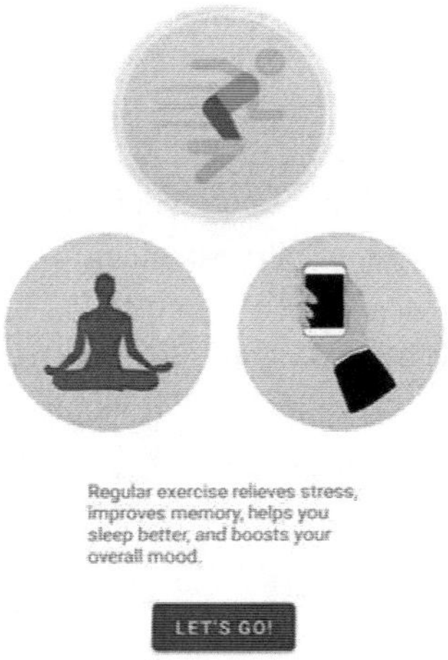

Figure 3.8: Recommended activities to the user

Figure 3.8 shows a screenshot of the application. The user can tap on any of the recommended activities and then page for that activity opens up and the user can start performing the activity. The cycle of recommending the activity and choosing is continued and performed every four hours.

4. **Results and Findings**

The benchmarks systems that we have considered to predict depression use twitter data. The benchmark system 0 used a Support Vector Machine classifier and achieved 70% accuracy in depression prediction.

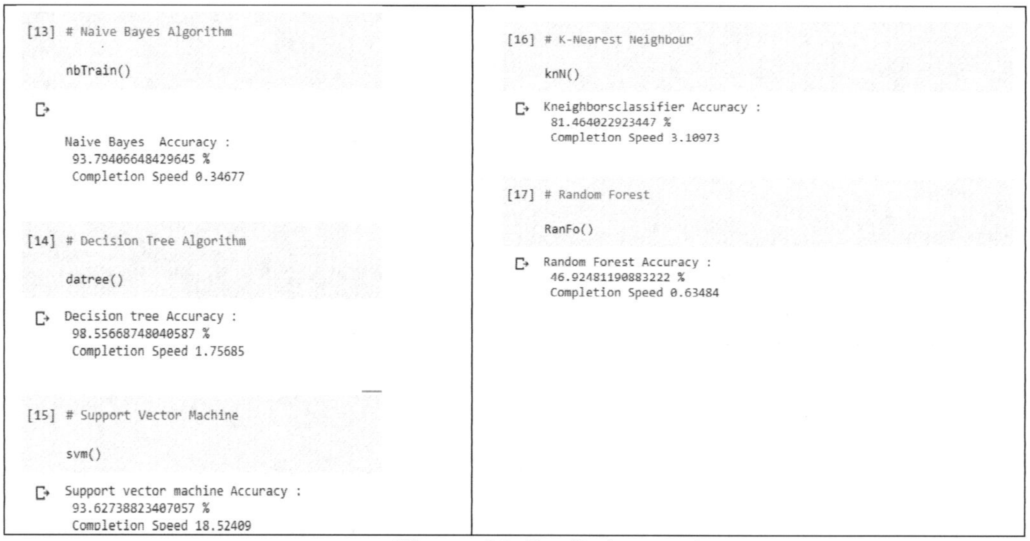

Figure 4.1. Prediction results

For depression prediction using patient history, we trained an Artificial Neural Network that gave an accuracy of 86.7% in 100 epochs. For depression prediction using social media data, we trained our model on five different classifiers. The Naïve Bayes classifier gave 93.79% accuracy, Support Vector Machine classifier gave an accuracy of 50.0%, k-Nearest Neighbour gave 81.46% accuracy, Random Forest gave an accuracy of 49.1% and Decision Tree classifier came out to be the best among all with an accuracy of 98.55%. Various results obtained by training our social media models are shown in Figure 4.1.

The confusion matrices for the different models trained for social media data is shown in Figure 4.2. The confusion matrix is a tabular representation of performance of a machine learning classification algorithm. These confusion matrices can be used to calculate the accuracy, precision and recall of a classifier. We can use the Sklearn library of the python to calculate these values. The columns represent the predicted label while the rows represent the true label for the posts. 1 represents positive post, 0 represents a neutral post while -1 represents a negative post. Each cell [i, j] (where i is the row and j is the column) in the matrix represents the number of posts with the true label represented by row i and the predicted label is represented using row j.

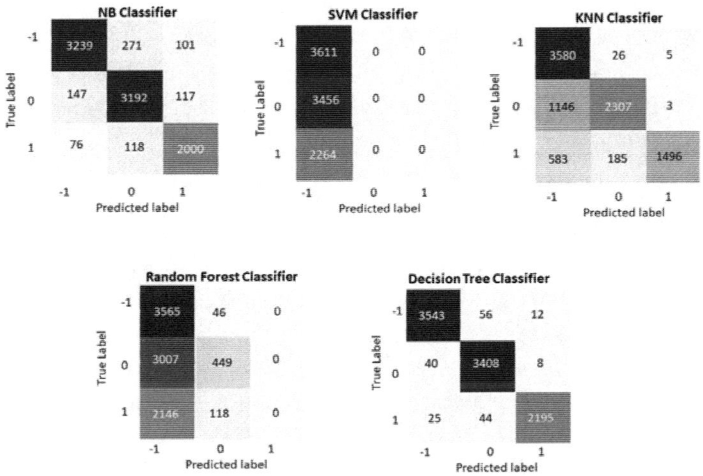

Figure 4.2. Confusion matrices

The confusion matrix for NB classifier tells that it classifies 2000 posts as positive sentiment posts correctly, while it wrongly predicted 118 positive posts as neutral posts and 76 posts as negative. The KNN classifier confusion matrix shows that it wrongly predicted 583 positive posts as negative and 185 positive posts as neutral posts. It predicted 1496 post correctly as having positive emotions. The confusion matrix of SVM classifier explains that it wrongly predicted 2264 positive posts as negative posts, and 0 posts as neutral & positive posts. The Random forest classifier correctly predicted 0 positive polarity posts as positive while 2146 positive posts as having negative polarity and 118 others as having neutral polarity. The Decision Tree classifier came out to be the best. Its confusion matrix says that it wrongly predicted 25 positive polarity posts as having negative emotion and 44 posts as neutral posts. It correctly predicted 2195 positive posts as having positive polarity better than all other classifiers.

5. Conclusion

Depression, which has become a global problem requires a cost-effective and accurate solution. We developed EmoCure mobile phone application that uses the idea of self-therapy to cure depression. The application is a step towards personalized patient-centric healthcare that is easily accessible and affordable. We use a unique combination of inputs that hasn't been used in previous research for prediction and recommendations. We successfully predicted depression using machine learning methods and performed personalised activity recommendations using smartphone usage patterns, wearable sensor data, social media data, and Electronic Health Records of the users. The prediction accuracy achieved and the effectiveness of the system in curing depression prove the efficacy of digital healthcare. Therefore, mental healthcare apps like EmoCure have the potential to treat and manage depression through activity recommendations.

6. Future Scope

The EmoCure app is only available for Android mobile phones currently. We would like to develop the same for the IOS platform too. In the future, we would like to improve the social media data analysis by using aspect-based sentiment analysis techniques. Enhancing social media data analysis will lead to more accurate predictions. We would also like to improve the activity recommended by recommending activities in accordance with the level of depression the patient is suffering from.

References

[1] WHO Publications, https://www.who.int/publications-detail/depression-global-health-estimates, last accessed 2020/4/13.Another reference

[2] Chaturvedi, S K, Rao, G P, Sarda, K D, & Suryawanshi, S Y 2014 Chronic pain and depression: an online survey on Indian experiences *Indian Journal of Pain* 28(3) 166.

[3] Aung, M H, Matthews, M, Choudhury, T 2017 Sensing behavioral symptoms of mental health and delivering personalized interventions using mobile technologies *Depression and anxiety* 34(7) 603-609.

[4] Number of Internet users in India from 2015 to 2018 with a forecast until 2023, https://www.statista.com/statistics/255146/number-of-Internet-users-in-india/, last accessed 2020/4/13.

[5] Number of facebook users in India from 2015 to 2018 with a forecast until 2023, https://www.statista.com/statistics/304827/number-of-facebook-users-in-india/, last accessed 2020/4/13.

[6] De Choudhury, M, Gamon, M, Counts, S, Horvitz, E 2013 Predicting depression via social media *Seventh international AAAI conference on weblogs and social media.*

[7] De Choudhury, M, Counts, S, & Horvitz, E 2013 Social media as a measurement tool of depression in populations *Proceedings of the 5th Annual ACM Web Science Conference* pp 47-56.

[8] Sheath, A, Jaimini, U, Thirunarayan, K, Banerjee, T 2017 Augmented personalized health: How smart data with iots and ai is about to change healthcare *IEEE 3rd International Forum on Research and Technologies for Society and Industry (RTSI)* pp 1-6.

[9] LiKamWa, R, Liu, Y, Lane, N D, Zhong, L 2013 Moodscope: Building a mood sensor from smartphone usage patterns *Proceeding of the 11th annual international conference on Mobile systems, applications, and services* pp 389-402 ACM.

[10] Hung, G C L, Yang, P C, Chang, C C, Chiang, J H, Chen, Y Y 2016 Predicting negative emotions based on mobile phone usage patterns: an exploratory study *JMIR research protocols* 5(3) e160.

[11] Eichstaedt, J C, Smith, R J, Merchant, R M, Ungar, L H, Crutchley, P, Preo?iuc-Pietro, D, Schwartz, H A 2018 Facebook language predicts depression in medical records *Proceedings of the National Academy of Sciences* 115(44) 11203-11208.

[12] Sonawane N., Padmane M, Suralkar V, Webel S, Date P 2018 Predicting depression levels using social media posts *International Journal of Innovative Research in Science, Engineering and Technology* **7** Issue 5.

[13] Islam, M R, Kabir, M A, Ahmed, A, Kamal, A R M, Wang, H, Ulhaq, A 2018 Depression detection from social network data using machine learning techniques *Health information science and systems* 6(1) 8.

[14] Gaur, M, Alambo, A, Sain, J P, Kursuncu, U, Thirunarayan, K, Kavuluru, R, Pathak, J 2019 Knowledge-aware assessment of severity of suicide risk for early intervention *The World Wide Web Conference* pp 514-525.

[15] Hung, G C L, Yang, P C, Wang, C Y, Chiang, J H 2015 A smartphone-based personalized activity recommender system for patients with depression *Proceedings of the 5th EAI International Conference on Wireless Mobile Communication and Healthcare* pp 253-257.

[16] Mohr, D C, Tomasino, K N, Lattie, E G, Palac, H L, Kwasny, M J, Weingardt, K, Caccamo, L 2017 IntelliCare: an eclectic, skills-based app suite for the treatment of depression and anxiety *Journal of medical Internet research* 19(1) e10.

[17] Yang, S, Zhou, P, Duan, K, Hossain, M S, Alhamid, M F 2018 emHealth: towards emotion health through

depression prediction and intelligent health recommender system *Mobile Networks and Applications* 23(2) 216-226.
[18] Wahle, F, Kowatsch, T, Fleisch, E, Rufer, M, Weidt, S 2016 Mobile sensing and support for people with depression: a pilot trial in the wild *JMIR mHealth and uHealth* 4(3) e111.
[19] https://github.com/niquejoe/Classification-of-Depression-on-Social-Media-Using-Text-Mining, last accessed 2020/05/05.
[20] https://zindi.africa/competitions/busara-mental-health-prediction-challenge/data, last accessed 2020/05/05.
[21] Bhatia S, Kesarwani Y, Basantani A, Jain S 2020 Engaging Smartphones and Social Data for Curing Depressive Disorders: An Overview and Survey *Paradigms of Computing, Communication and Data Sciences (PCCDS).*
[22] Dalal S, Jain S, Dave M 2019 A Systematic Review of Smart Mental Healthcare, *Proceedings of the 5th International Conference on Cyber Security and Privacy* (ICCS) pp 38-44.
[23] Mishra S, Jain S, Abraham A, Shandilya S Dec 2018 Secure Semantic Smart HealthCare (S3HC) *Journal of Web Engineering, River Publishers* **17:8** pp 617-646.

A survey on data integrity verification schemes using blockchain technology in Cloud Computing Environment

K GANGADEVI [1] and R RENUGA DEVI [2]

[1,2] Department of Computer Science, VELS University, Pallavaram, Chennai, India
[1]seenuganga@yahoo.co.in and [2] nicrdevi@gmail.com

Abstract. Cloud computing is the delivery of different services for business or any organization to store their data and utilize the various resources over the internet. The resources of cloud computing is storage, software, computing hardware and services. Usefulness of the cloud computing infrastructure is to grant services based on demand to the user. The security is a major concern in the cloud computing that are many modes of securities including accessibility, authentication, data integrity, data recovery and confidentiality. The data integrity is major role in the cloud computing security and storage service. Hence, the data integrity scheme provides the proof the truthfulness of the stored data in the cloud. In this paper, we are conducting a brief survey on data integrity verification schemes for cloud computing based on blockchain technology. The aim of this survey is to enhance the understanding the security issues and discuss the importance of data integrity scheme for outsourced data.

Keywords: cloud computing, data integrity, proof data possessions, proof of retrievability, blockchain technology, validation.

1.Introductions

The cloud computing is gained tremendous popularity due to outsourcing computation and providing storage requirements based on the user availability. The user can use the resources over the internet such as storages, services, applications, and bandwidth based on distributed shared services. The resource provider is called the cloud service provider (CSP). The cloud computing is organized into three categories based on the services. (i) Infrastructure-as-a-Service (IaaS), (ii) Platform-as-a-Service (PaaS) and (iii) Software-as-a-Service (SaaS). Three deployment models in cloud environment are private, public and Multi or Hybrid models [1]. One primary feature of cloud computing is the outsourcing the user data in cloud environment with centralized manner. The services of outsourced storage in cloud environment have an innovative yield development point by providing a comparable high scalable, low-cost, and location-independent environment for the control the data. Many cloud based storages are available today such as One Drive, Amazon, Google Drive and Drop Box.

A cloud storage service (CSS) is a critical cloud computing services that belongs to IaaS. The CSS is to mitigate the load of storage management and preservation of local storage. However, this kind of significant service affects by malicious attacks or failures both from outside and inside the cloud,

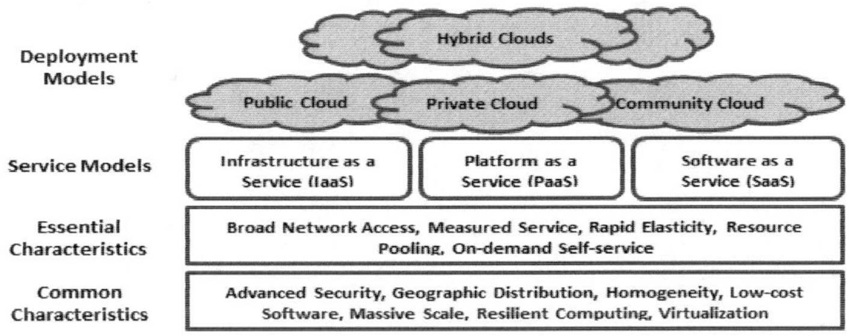

Figure 1.1: Architecture of cloud environments

modification, corruption and deletions. This may be caused by careless scheme preserve and disturb the cost saving behaviour. Hence, CSP required allowing an efficient auditing service to validate the integrity and accessibility of the stored data. The CSP is enclosed by a service level agreement (SLA) to make sure the security of the outsourced data [2]. The architecture of cloud computing environment is shown in Figure 1.1.
The regular solution for make sure the integrity of outsourced data is to utilize conventional cryptographic models based on signature and hashing methods. But, discussed conventional methods need a local copy of data to make sure the integrity of outsourced data [3]. This would be reached an expensive I/O cost and massive communication overhead across the networks. In the recent years, the verifying the outsourced data integrity n cloud storage is more attention for industry persons and researchers. The aim of this paper is conducting a brief survey on data integrity scheme based on blockchain technology. First, an introduction and fundament details of data integrity scheme, fundamental details of blockchain technology, various data integrity scheme for verifying outsourced data in cloud based on blockchain technology schemes. Finally, present the various challenges of blockchain technology when apply to data integrity scheme.

1.1 Data integrity

Data integrity is the guarantee that the digital information is uncorrupted and can be accessed or modified by approved users. The data integrity involves preserving the stability, accurateness and reliability of data. The integrity verification scheme is a significantly essential for data assurance and digital forensics. The data integrity verification process is a major task in cloud computing because of the data storage is location–independent behaviour [4]. In the earlier stage, the signature scheme and hash functions are used to check data integrity. This both scheme is want to download complete data from the cloud storage then hash value or signature value of download data are compared with local data. So, the hash tree is resolving the above mentioned fault, but it only validate the exactness of the hash tree corresponding to the cloud storage rather than the data itself. Hence, the designing the significant mechanism is to verify the data integrity significantly without having to recover the complete data.

IOP Conf. Series: Materials Science and Engineering 1110 (2021) 012011 doi:10.1088/1757-899X/1110/1/012011

To solve abovementioned shortcomings, many research works have been conducted on allowing public auditing scheme by employing data integrity verification methods. Generally, the data integrity is an interactive protocol running between CSSs and user. The data integrity scheme performed between CSS and user is called private auditing. In the private auditing scheme, the users required to calculate some authentication data internally and often communicate with a CSS during storing the data, which takes more computation burden.

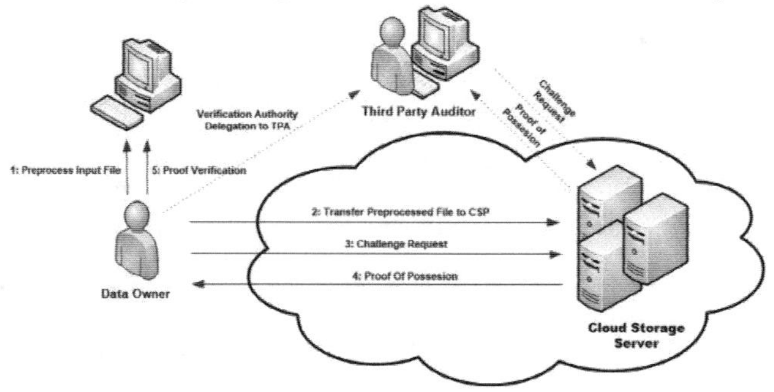

Figure 1.2: Architecture of data integrity scheme

The third-party auditor (TPA) is launch to communicate with user for auditing. The auditor can be either TPA or authentic user is called public auditing. The TPA is used to improve auditing effectiveness and reduce the computation cost. Consequently, developing a proper data integrity verification model for public auditing is considered as a vital requirement [5].

Two major types of data integrity schemes are used for public auditablity of the outsourced data, such as Provable Data Possession (PDP) [6] and Proof of Retrievability (PoR) [7]. PDP scheme was introduced by Ateniese et al. (2007) [6]. In the PDP scheme, the client is sent the pre-processed the data to an untrusted server for storage. Also, it is keeping a small amount of meta-data. Later, the client requests the server to demonstrate that the stored information has not been corrupt with or deleted without downloading the authentic data. However, if the data once corrupted or deleted by the malicious user, the PDP scheme is not retrieved properly. This behavior of PDP has many short comings including data loss, financial scratch, and loss of trust. PoP is another data integrity scheme similar to that of PDP [7]. It has another advantage contrast with PDP that will recover the loosed data or corrupted data by using error correcting code (ECC) method.

The data integrity can be classified into two categories either deterministic or probabilistic. The deterministic scheme is necessary to access the entire file in order to find out the integrity [8]. This scheme is not suitable for large files because it will take more time for completing the process of integrity verification. The probabilistic scheme is dynamically accessing the data block to verify the data integrity [1]. This scheme is more suitable for large files to complete with less computational time. The performance of this scheme is poor when size of data is low. The nature of data is significant role in the data integrity verification process. The data can be classified based on its behavior either static nature or dynamic nature. The static nature of the data cannot modify once stored in the cloud. Static nature of the data is having small storage overhead and given that higher acceptance speed for errors [9]. In term of dynamic nature, the data can be insertion, updation, deletion and modification in the cloud storage [10].

1.2 Architecture of data integrity scheme

The architecture of the data integrity scheme for outsourced data is discussed in this section and architecture of data integrity scheme is portrayed in Figure 2. The cloud data storage services consist of four phases which is explained as follows,

- *Data Owner (DO):* Who is utilizes the CSS for storing their huge amount of data
- *Cloud Service Provider (CSP):* who is providing the CSS and computational resources
- *Third Party Auditor (TPA):* she/he has capabilities to supervise the outsourced data under the allocation of data

The DO generally adopts a TPA to accomplish the data integrity verification procedure to evaluate the integrity of outsourced data. DO can't be entirely authenticated the consistency of the TPA during the integrity verification process. Hence, furnish over the verification of data integrity to the malicious TPA may lead to threats of data security. In the earlier stage, the data integrity scheme is performed on the static data and then extended to dynamic data. Similarly, the private data integrity verification scheme is conducted in the earlier stage and then public data integrity scheme is launched to verify the data with the help of TPA or without.

2.Blockchain technology

Recently, cryptocurrency has concerned widespread attention from both academic and industries. The bitcoin is a first crytocurrency and the core mechanism of blockchain technology. The blockchain is introduced by Nakamoto (2008) [11]. Blockchain is decentralized digital public ledger for distributed storage and peer-to-peer communication between nodes. Blockchain is facilitated by combining with numerous core technologies including digital signature, cryptographic hash, time stamp and distributed consensus algorithm. In blockchain, all committed transactions are stored in the form of chain of blocks. Chain of blocks is continuously increased one by one when the new blocks are created [12]. The blockchain technology is more efficient and cost effective because of transactions can store in the decentralized manner. The blockchain can be applied various fields of application including cloud computing [13], big data environment [14], financial service [15] and Internet of Things (IoT) [16].

2.1 Architecture of blockchain technology

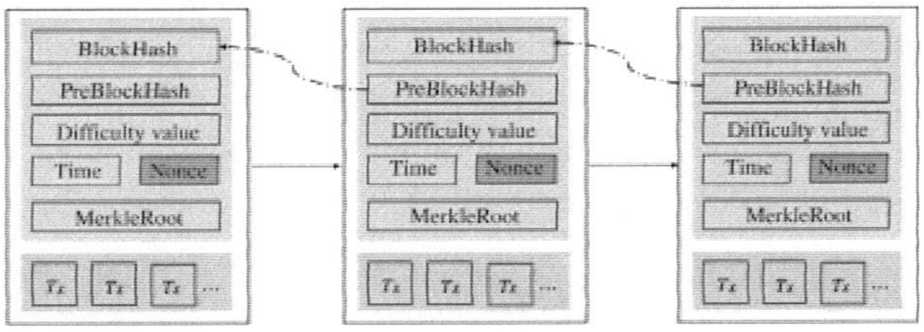

Figure 1.3: Architecture of Blockchain Technology

The blockchain is an order of blocks that holds an entire list of transaction records. The architecture of blockchain shows in Figure 1.3. Each block is a directly point to the previous block using hash value that the prior block is called parent block. The first block of blockchain is called genesis block and it has no parent block. Each block of blockchain has two major components including a block header and a block body. The block header contains much information such as block version, parent hash values, Markel root hash, time stamp, nBits and nonce [17]. The block body contains a detailed transaction data and a transaction counter.
The maximum numbers of transactions are determined based on the block size and transaction. The blockchain uses method of an asymmetric cryptography to validate the transactions. A digital signature based on asymmetric cryptography is used in an unreliable environment. Each user holds either a private key or a public key. The use of private key is sign the transactions. The signed transactions are extended during the entire network that accessed by public key. The digital signature is classified into two phases including signing and verification phases. In the blockchain architecture, the elliptic curve digital signature algorithm (ECDSA) is used as a classic signature [18].

2.2 Types of blockchain technology
Three types of the blockchain technologies are private, public and consortium blockchain [19] which is represented in Figure 4.

2.2.1 Private blockchain
The private blockchain is a distributed database. This approach favours (i) by allowing few organizations with privacy requirements to implement blockchain; (ii) cryptographic audit adds value. The permissions are centralized to an organization. It is an easier method due to the small number of participating users. Hence, the transaction and processing time is low. It is an inverse of the public blockchain due to environment friendly as less computational power is required. Comparing to the public blockchain, it is less secure.

2.2.2 Public blockchain
The public blockchain allows any user to add, download data or block in the blockchain network. The public blockchain is a release for all kinds of entities to contribute in the networks. It is a completely transparent and decentralized database of the transaction like an open network. It is highly secured using cryptography and consensus protocol which is an answer for those who criticize that it's not secured since the public blockchain is an open source and low privacy. It is less eco-friendly due to high computational power and energy is required.

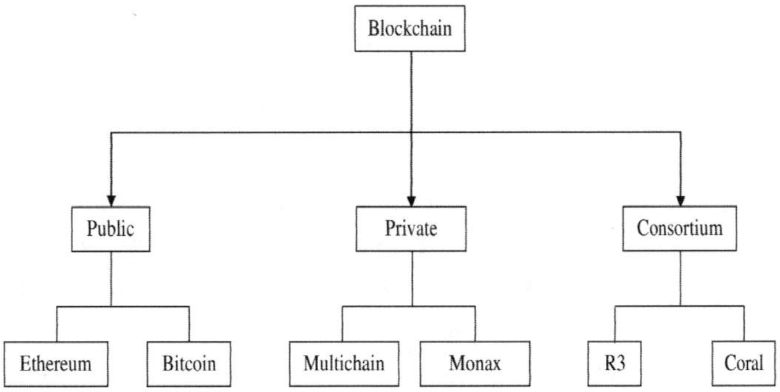

Figure 2.1: Types of blockchain technology

2.2.3 Consortium blockchain

The consortium blockchain is "partially private" works across different organizations with pre-defined set of users. It is a fast, delivery with better scalability and privacy.

Algorithms	Energy Consumptions	Transactions/Sec	Computation cost	Example
PBFT	High	-	Low	Hyberledger
PoW	High	7-30	High	Bitcoin
PoS	Low	30-173	Low	Peercoin
DPoS	Very low	25-2000	Low	Bitshares
Ripple	High	-	Low	-
Tendermint	Low	-	Low	

Table 1: Comparison of consensus algorithms

2.3 Consensus algorithms

how to achieve consensus mechanism among the undependable nodes is a transformation of the problem of Byzantine Generals (BG) [20]. In blockchain like a BG problem, there is no middle node. Nodes are no need to trust each other. Hence, some set of rules are required to make sure that the ledgers in different nodes are reliable [21]. Consensus algorithm is a mechanism to make sure the precision and reliability of information stored by all nodes in a distributed ledger. Consensus mechanism is used to authenticate every single block and eventually used to achieve the required consensus. Also, it is a process of removing invalid or contradictory transactions. The many consensus algorithms are used in blockchain technologies which are discussed in the following sub paragraphs. Table 1 illustrate the comparisons of the consensus algorithms.

2.3.1 Practical byzantine fault tolerance (PBFT)

PBFT is a replication algorithm which can handle the problem of byzantine problem [20]. Hyperledger fabric is used in PBFT as a consensus algorithm and it can be handled 2/3 malicious byzantine replica.

A new block is resolute in a round and it is a most important that chosen depends upon the some policy. The policy is responsible for ordering the transactions. In the PBFT, the entire process is separated into three parts such as pre-prepared, prepared and commit.

2.3.2 Proof-of-Work (PoW)

The PoW is a well-known consensus mechanism used in Bitcoin [11]. Each node in the network is computing a hash value of the altering block header. The consensus mechanism is wants that the calculated values must smaller or equal to the given value. All participants in the network have to compute a hash value using different nonce value until reach the target. If anyone node achieve relevant value, then all others must confirmed the correctness of the value. The node calculated the hash values is called miner and procedure of PoW is called mining. However, it required a complicated computational process in the authentication process.

2.3.3 Proof-of-Stack (PoS)

PoS is take power consumption and enhance the node performance when compared with PoW [22]. PoS require validating the ownership of the amount of currency instead of trying client to find a nonce in an infinite gap. The PoS protocol randomly provides privilege to create a block between the selected validators based upon the value of their stakes. The selected validator rewarded by the transaction fee partially or entirely. But, it produces inaccuracy during vulnerable attacking. Compared with PoW, PoS has low latency, lower resource usage and computation power.

2.3.4 Delegated PoS (DPoS)

DPoS is a consensus algorithm upholding an unquestionable contract on the fact across the network and transaction validations. Among the validator, the voting power is to determine the peer who fills the role of a delegate for maintaining the network and transaction validations known as forging. The DPoS is to hold stake vote to select by block verifiers or creators.

2.3.5 Ripple

Ripple is a kind of consensus algorithm that employ collectively trusted and sub-networks within the bigger networks [23]. The participating nodes are separated into two types such as server and client. The server performs the contribute process of consensus. The client performs only transferring of funds. In ripple, each server has a unique node list (UNL) that is a significant value to the server to query. The server would query the node in UNL when put a transaction in the ledger. The transaction is crammed into the ledger if the received agreement has reached 80%. The fault node is corrected by the ledger whose the percentage has less than 20 %.

2.3.6 Tendermint

Tendermint is a kinds of byzantine consensus algorithm. In which, a new block is resolute in a round manner [24]. A proposer would be chosen to relay an unverified block. Hence, all nodes in the network are required to know the proposed selection. Tendermint could divide into three steps prevote step, pre-commit step and commit step. The validator decide whether to relay a prevote for the planned block in the prevote step. The pre-commit step is performed if the node has obtained more than 2/3 of the planed block, then it relay a pre-commit for that block. If the node has obtained over 2/3 of pre-commits, it goes into the commit steps.

2.4 Characteristic of blockchain technology

The blockchain technology is a decentralized ledger scheme with high security to enhance the integrity and trust of transactions [25]. This section discusses some major characteristics of blockchain as follows,

- *Decentralized:* the decentralized property of the blockchain can manage the single point of failure and scalability. The property of decentralized and distributed ledger in blockchain is reducing the latency.
- *Transparent:* the blockchain technology is produced high level transparency with the help of sharing transaction information between all participants' users who involving the transactions. The role of a third party is no need to verify the any transaction as in the centralized technology.
- *Immutability:* the blockchain can ensure the every transaction by creating the immutability ledger. In the blockchain, the every transaction in the block is cannot change. Because, the blocks of the every transactions are stored permanently by connecting each other using hash values.
- *Security:* it provides more security for storing the data by using public key infrastructures and protect from changing by the malicious user. The blockchain mechanism protects the entire system by eliminating from the problem single point of failure.
- *Efficiency:* the distributed mechanism is used to enhance the efficiency which verifies the all transaction in the distributed environments. The blockchain is more efficiently compared with a centralized system in terms of completion speed, cost and risk management.
- *Traceability:* the property of timestamps in the blockchain is used to recognize and store each transaction. This helps in participating node to keep the order of transactions for data traceable. The timestamp property can help to reduce the cost of transaction traceability. The traceability of the blockchain system is secure and consistent, and solves the problem of double spending.
- *Anonymity:* The blockchain technology is using the asymmetric encryption techniques for encrypting the data. The encryption of data in the blockchain is used two methodologies including data encryption and digital signatures. Data encryptions are used to make sure the security of transaction data and decrease the possibility of losing transaction data. Transaction data is use the digital signed to indicate the identity of the originality of transaction data.

3. Blockchain based data integrity scheme:

The cloud storage is widely used resource service compared with other resource due to its higher efficiency and low cost. The cloud user can enjoy high-quality storage service for their data. Because the user data store is resides at isolated server. The security issue of CSS is a major role that makes the user worried about their outsourced data. In the earlier stage, the many public data integrity scheme is depending upon a trusted TPA to ensure the integrity of the data.

However, the malicious TPA can collaborate with the CSP to obtain various benefits. In order to

Table 2 : Performance comparisons of data integrity schemes

Ref. No.	Batch	Dynamic	Public	Private	Block less	Certificate less
[16]	✓		✓			
[17]	✓		✓			
[26]			✓			✓
[28]	✓	✓	✓			
[30]				✓		
[31]			✓			
[32]			✓			✓
[33]		✓	✓	✓		
[34]		✓	✓			
[35]		✓	✓			
[36]						
[37]			✓			
[38]		✓	✓			
[39]			✓		✓	
[40]			✓		✓	

mitigate the user tension, the data integrity scheme is very important to verify the integrity of the outsourced data [26]. Hence, the auditabilty of property of the blockchain is utilized for data integrity checking that the transaction records of CSS can be easily verified and tracked [2]. The blockchain can successfully preserve the data consistency in the distributed computing systems [27]. Also, the blockchain technology is used for verifying the data integrity of stored data in different platforms such as Big data [28], IoT [29], cloud environment [30]. The following paragraphs are discussed some research papers about data integrity inspection of the cloud storage data by using blockchain technology.

Xue et al. (2019) [17] propose an identity based public auditing scheme by using blockchain for maintaining data integrity of outsourced data against various attacks. The authors use the nonce of public blockchain (Bitcoin) for selecting the challenge message of identity- based public auditing (IBPA). Additionally, the auditing results are stored in the public blockchain with traceable and auditable. Security of IBPA is under the oracle model based on Diffie-Hellman problem. The proposed data integrity scheme is produced low communication overhead and low computational cost. Yu et al. (2017) [28] developed a decentralized big data auditing method for the applications of smart city using ability of blockchain to enhanced reliability and strength without the need of a centralized TPA. Wang et al. (2019) [30] developed blockchain-based private PDP data integrity scheme. In which, the data irreversibility property of the blockchain is used to store some metadata for client that can be used to verify the remote data integrity. The proposed method uses the blockchain that produce the corresponding metadata and forwarded it to the client.

Wei et al. (2020) [31] proposes the data integrity protection scheme by using blockchain technology. The developed data integrity scheme is built by virtual machine proxy method and unique hash value corresponding to the file produced by the merkal hash tree (MHT) which is used to examine the changes of files. Zhang et al. (2019) [32] develop certificateless public verification method in opposition to procrastinating auditors (CPVPA) by means of blockchain technology. The key thought is the auditors need to store each verification outcome in a blockchain. Because, the transactions of the blocks are time-sensitive, the verification process can be time-stamped after the related transaction is stored. It can use to verify whether auditors achieve the verifications at the approved time.

Hao et al. (2019) [33] propose a decentralized form based on blockchain that contains the several collaborative verification peers (VPs). The each VPs maintains a replication of the complete blockchain to keep away from malicious tampering. The proposed scheme is allow DO to store and make sure the verification information by writing and get back the blockchain. Also, proposed scheme is extend the algorithm by introducing then verification group (VG) constituting by some VPs controlled by Inter-Group and Inner-Group consensus protocols. Lu et al. (2020) [34] design significant decentralized data integrity auditing technique based on Hyperledger Fabric (HF-Audit). TPA can be dynamically determined for each auditing task by using Hyperledger Fabric as a communication platform. Also, the proposed system design a significant auditing method for data integrity based on bilinear pairing and commitments in order to enhance the scalability.

Wang et al. (2020) [35] proposed a blockchain-based fair payment that adopting blockchain to replace TPA for public cloud storage auditing. In the proposed system, DO and CSP will run a blockchain-based smart contract to ensure the details of transactions. The CSP is necessary to submit data possession proof frequently. The proposed scheme is present no interactive public PDP and design a smart contract using blockchain technology for public cloud storage auditing. Kalis et al. (2018) [36] develop a hash validation technique using blockchain technology. The actual stored data can be authenticated against the blockchain based hash and an employ inside an application check trail to authenticate the audit trail data. This implementation demonstrates that the hash validation is capable to identify malicious and fortuitous changes that were prepared for the data.

Wang et al. (2019) [37] develop a new personal health record sharing techniques with data integrity verification method based on blockchain technology. The proposed method uses the attribute-based encryption (AE) and searchable symmetric encryption (SE) techniques to accomplish privacy protection, fine-grained access control and keyword search. The blockchain based method is used to supervise propose a bilinear mapping and blockchain based data integrity scheme (BB-DIS) for large-scale IoT data to reduce computational and communication overhead. In the proposed method, large scale IoT data are segmented into

Table 3 : Comparisons of the data integrity scheme

Ref. No.	Verification Algorithm	Consensus algorithm	Merits	Demerits
[16]	Lifted EC-ElGamal cryptosystem, bilinear pairing	PoW	High accuracy	High computation cost
[17]	Diffie-hellman	Bitcoin system	Low communication overhead and computational cost	Lack of scalability
[28]	Bilinear pairings	PBFT	lower communication and computation costs	Low accuracy
[30]	RSA	Blockchain	More efficient and practical	High computational time
[31]	-	PoW	Produced warning message	Energy consumption is high
[32]	Bilinear maps	PoW	Avoid the certificate management problem with high accuracy	More Energy consumptions
[33]	RSA	VG-WriteBlock. VG-CheckBlock.	High security and efficiency	Energy Consumptions
[34]	bilinear pairing and commitments	HF-Audit	High security and scalability	High energy consumptions
[35]	Bilinear Pairings + Diffie-Hellman	PoW	Low execution time	High computational burden
[36]	-	PoC based hash validation function	Can able to distinguish tampering	It cannot prevent tampering
[37]	SE+AE	Ethereum	avoiding the single point failure problem	High computation overhead
[38]	Bilinear Mapping	PBFT	Computational cost and communication overhead	Inability to handle complex data types
[39]	Merkle tree + sampling strategy	Ethereum	Enhancing the performance of data integrity verification scheme.	More power consumptions
[38]	Merkle tree + sampling Strategy with random generation	Ethereum	More flexibility, transparent and auditable	More power consumptions

shards and homomorphic verifiable tags (HVTs) are produce for sampling confirmation. Then, data integrity method can be accomplished according to the individuality of bilinear mapping and blockchain transactions.

Yue et al. [39] proposes new data integrity checking scheme based blockchain in Peer-to-Peer cloud storage. The present Merkle tree for data integrity checking and investigate the performance of system under dissimilar Merkle trees construction. Also, the authors develop a logical sampling scheme to create sampling checking more efficient. Yuea et al. (2020) [40] propose a new data integrity checking scheme in a decentralized edge-cloud storage (ECS) using blockchain-based framework without TPA. The authors employ the Merkle tree with dynamic demanding records to verify data integrity and optimize the system performance by analysing different Merkle tree structures. Zhao et al. (2020) [16] designed a new remote data integrity verification technique for security of cloud-based information systems. The proposed scheme is combined with leverages the Lifted EC-ElGamal cryptosystem, bilinear mapping and blockchain scheme to carry out proficient public batch signature checking, keep the security and data privacy of the IoT environment. The comparisons of reviewed research paper are shown in Table 2 and 3.

4. Limitations of blockchain based data integrity scheme

DO combine with TPA to achieve the data integrity checking scheme to assess the outsourced data integrity. During the verification process, DO can't be completely authenticated the consistency of the TPA. Hence, furnish over the checking of data integrity to the malicious and it may direct to data security threats. Table -3 is demonstrates the advantages and disadvantages of existing research papers for problem of data integrity. However, the blockchain based data integrity scheme has been produced higher accuracy when compared with conventional verification scheme. But, the existing research works have to look the troubles of huge computational and communication overhead [38]. The various research works have been developed a blockchain based verification scheme such as [17] , [31] and [35]. In the blockchain, the bitcoin is used PoW consensus protocol to verify the integrity of outsourced data. However, the PoW is taking more energy consumption and consensus process is prohibitively slow for validating [41].

5. Challenges of data integrity verification methods

The data integrity scheme has many challenged during integrity verification using blockchain technology including computation, storage, communication and disk I/O efficiency. The following subsections are discussed about the challenges of data integrity scheme.

- *Computational efficiency:* the pre-processing step is a major concern when data outsourced in the cloud. Proof generation of outsourced data during the data integrity process is taking more computation efficiency. Also, taking more computation efficiency when dealing with meta data.
- *Storage efficiency:* the additional meta data is required along with original data for integrity verification and data recovery. Hence, in this circumstance, the storage efficiency is considered as a very important.
- *Communication efficiency:* Communication effectiveness of the data integrity scheme is considered in three ways such as the original data transfer to the cloud with meta data is overhead, request to the verification process by the data owners, and proof of data integrity is sent reverse to the CSS.
- *Disk I/O efficiency :* Accessing the block for making proof , inconsistent length of block access , amount of blocks , and unordered block indexes are the disturb the efficiency of the data integrity scheme.

6. Conclusions

Data integrity and confidentiality are the major security concerns in a cloud computing environment. Many data integrity models have been developed over the years to concentrate on the data integrity issues. On the other hand, the blockchain technology is a well-known and considerable technique in term of security issues in the various fields. The blockchain technology helps to enhance the performance of data integrity models. The major focus of the paper is to enhance the perception of security issues of data integrity methods based on blockchain technology. The present paper gives a comparative analysis based on characteristics of data integrity schemes and blockchain technology. Also, it recognized the timeline of the development of data integrity schemes based on blockchain technology. Also, it can be identified the future aspects of blockchain based data integrity scheme. This paper can help to researchers to carry out their future research work. Further, we identify the challenges and security issues of data integrity schemes based on blockchain technology.

7. Future enhancements

In the data integrity scheme, the verification process is significantly essential process. The many research work have been applied a blockchain technology for data integrity verification. In this circumstance, the enhancing performance of the blockchain validation process is challenging and crucial task. Hence, the PoS, DPoS is an alternative consensus algorithm for validating the transaction of blockchain technology. On the other hand, PeerCensus is a well-known consensus algorithm proposed by Decker et al. (2016) [41] that proceed as a certification influence and supervises peer identities in a P2P network, and ultimately improves performance of blockchain. As well, Kraft (2016) proposes a new consensus algorithm that enhance stability of bitcoin performance [42].

References

[1] Nepal S Chen S Yao J and Thilakanathan D 2011 DIaaS: Data integrity as a service in the cloud IEEE 4th International Conference on Cloud Computing, IEEE.

[2] Fan Y Lin X Tan G Zhang Y Dong W and Lei J One secure data integrity verification scheme for cloud storage *Future Generation Computer Systems* **96** 376-385.

[3] Sood S K 2012 A combined approach to ensure data security in cloud computing *Journal of Network and Computer Applications* **35** 1831-1838

[4] Shen W Qin J Yu J Hao R Hu J and Ma J 2019 Data integrity auditing without private key storage for secure cloud storage *IEEE Transactions on Cloud Computing*.

[5] Zhu Y Hu H Ahn G-J and Yau S S 2012 Efficient audit service outsourcing for data integrity in clouds *Journal of Systems and Software* **85** 1083-1095

[6] Ateniese G Burns R Curtmola R Herring J Kissner L Peterson Z and Song D 2007 Provable data possession at untrusted stores Proceedings of the 14th ACM conference on Computer and communications security

[7] Juels A and Kaliski Jr B S 2007 PORs: Proofs of retrievability for large files Proceedings of the 14th ACM conference on Computer and communications security

[8] Han S Liu S Chen K and Gu D 2014 Proofs of retrievability based on MRD codes International Conference on Information Security Practice and Experience, Springer.

[9] Yuan J and Yu S 2013 Proofs of retrievability with public verifiability and constant communication cost in cloud Proceedings of the 2013 international workshop on Security in cloud computing Editior.

[10] Li A Tan S and Jia Y 2019 A method for achieving provable data integrity in cloud computing *The Journal of Supercomputing* **75** 92-108.

[11] Nakamoto S. 2008 Bitcoin: A peer-to-peer electronic cash system.(2008) Editior.

[12] Casino F Dasaklis T K and Patsakis C 2019 A systematic literature review of blockchain-based applications current status classification and open issues *Telematics and Informatics* **36** 55-81

[13] Yang C Chen X and Xiang Y 2018 Blockchain-based publicly verifiable data deletion scheme for cloud storage *Journal of Network and Computer Applications* **103** 185-193.

[14] Abdullah N Hakansson A and Moradian E 2017 Blockchain based approach to enhance big data authentication in distributed environment 2017 Ninth International Conference on Ubiquitous and Future Networks (ICUFN) IEEE.

[15] Zhang L Xie Y Zheng Y Xue W Zheng X and Xu X 2020 The challenges and countermeasures of blockchain in finance and economics *Systems Research and Behavioral Science* **37** 691-698.

[16] Zhao Q Chen S Liu Z Baker T and Zhang Y 2020 Blockchain-based privacy-preserving remote data integrity checking scheme for IoT information systems *Information Processing & Management* **57** 102355

[17] Xue J Xu C Zhao J and Ma J 2019 Identity-based public auditing for cloud storage systems against malicious auditors via blockchain *Science China Information Sciences* **62** 32104.

[18] Johnson D Menezes A and Vanstone S 2001 The elliptic curve digital signature algorithm (ECDSA) *International journal of information security* **1** 36-63.

[19] Atlam H F and Wills G B 2018 Technical aspects of blockchain and IoT

[20] Lamport L Shostak R and Pease M 1982 The Byzantine Generals Problem ACM Transactions on Progamming Languages and Syetems vol 4 no 3 pp. 382-401

[21] Li X Jiang P Chen T Luo X and Wen Q 2020 A survey on the security of blockchain systems *Future Generation Computer Systems* **107** 841-853.

[22] Zamfir V 2015 Introducing casper the friendly ghost *Ethereum Blog URL: https://blog. ethereum. org/2015/08/01/introducing-casperfriendly-ghost.*

[23] Schwartz D Youngs N and Britto A 2014 The ripple protocol consensus algorithm *Ripple Labs Inc White Paper* **5**

[24] Kwon J 2014 Tendermint: Consensus without mining *Draft v. 0.6, fall* **1**.

[25] Atlam H F and Wills G B 2019 Technical aspects of blockchain and IoT Advances in Computers Elsevier.

[26] Saxena R and Dey S 2019 Data integrity verification: a novel approach for cloud computing *Sādhanā* **44**, 74.

[27] Zhu Z Qi G Zheng M Sun J and Chai Y 2019 Blockchain based consensus checking in decentralized cloud storage *Simulation Modelling Practice and Theory* 101987.

[28] Yu H Yang Z and Sinnott R O 2018 Decentralized big data auditing for smart city environments leveraging blockchain technology *IEEE Access* **7** 6288-6296

[29] Machado C and Fröhlich A A M 2018 IoT data integrity verification for cyber-physical systems using blockchain," in 2018 IEEE 21st International Symposium on Real-Time Distributed Computing (ISORC), IEEE.

[30] Wang H Wang Q and He D 2019 Blockchain-Based Private Provable Data Possession *IEEE Transactions on Dependable and Secure Computing*.

[31] Wei P Wang D Zhao Y Tyagi S K S and Kumar N 2020 Blockchain data-based cloud data integrity protection mechanism *Future Generation Computer Systems* **102** 902-911

[32] Zhang Y Xu C Lin X and Shen X S 2019 Blockchain-based public integrity verification for cloud storage against procrastinating auditors *IEEE Transactions on Cloud Computing*.

[33] Hao K Xin J Wang Z and Wang G 2020 Outsourced data integrity verification based on blockchain in untrusted environment *World Wide Web*, 1-24.

[34] Lu N Zhang Y Shi W Kumari S and Choo K-K R 2020 A secure and scalable data integrity auditing scheme based on hyperledger fabric *Computers & Security* **92** 101741

[35] Wang H Qin H Zhao M Wei X Shen H and Susilo W 2020 Blockchain-based fair payment smart contract for public cloud storage auditing *Information Sciences* **519** 348-362.

[36] Kalis R and Belloum A 2018 Validating data integrity with blockchain 2018 IEEE International Conference on Cloud Computing Technology and Science IEEE

[37] Wang S Zhang D and Zhang Y 2019 Blockchain-Based Personal Health Records Sharing Scheme With Data Integrity Verifiable *IEEE Access* **7** 102887-102901.

[38] Wang H and Zhang J 2019 Blockchain Based Data Integrity Verification for Large-Scale IoT Data *IEEE Access* **7** 164996-165006

[39] Yue D Li R Zhang Y Tian W and Peng C 2018 Blockchain based data integrity verification in P2P cloud storage IEEE 24th International Conference on Parallel and Distributed Systems (ICPADS), IEEE.

[40] Yue D Li R Zhang Y Tian W and Huang Y 2020 Blockchain-based verification framework for data integrity in edge-cloud storage *Journal of Parallel and Distributed Computing*

[41] Decker C Seidel J and Wattenhofer R 2016 Bitcoin meets strong consistency Proceedings of the 17th International Conference on Distributed Computing and Networking Editior

[42] Kraft D 2016 Difficulty control for blockchain-based consensus systems *Peer-to-Peer Networking and Applications* **9** 397-413.

Multi-Agents Based Coordinated Scheduling in Hospitals

J.Dhiviya Rose[1] **and S.Christalin Nelson**[2]

[1] School of Computer Science, University of Petroleum and Energy Studies (UPES) Bidholi, Dehradun. INDIA 248007
[2] School of Computer Science, University of Petroleum and Energy Studies (UPES) Bidholi, Dehradun. INDIA 248007

Email: dhiviyanelson@gmail.com

Abstract. — Patient Scheduling is the process of scheduling the patients by creating an optimal schedule for the patients in using the resources in hospitals. The real time scheduling with various constraints like emergency situations like accident case makes this problem complex. The manually scheduling algorithms and the traditional approaches needs more coordination, time and manpower to achieve its efficiency. Agents are a natural choice because they are known to represent a distributed environment well and they are capable of coordinating with each other in dynamic environment. The architectural components like Belief that maintains the internal state of the system, Desires that represents the motivational state of the agent and Intentions that helps to deliberate the state of the goal make this autonomous system adaptable and fault tolerant. In the proposed system each resources and patient is constructed to be software agents, which has the ability to coordinate among them. The optimal schedule is calculated in minimal time using the Genetic Algorithm (GA), an efficient search-based optimization technique. When a patient arrives, he/she is served by a common agent where the priority, resource requirement, processing time, and sequence are assigned to each patient. Hence in this highly dynamic domain optimal schedule generation with minimal time and hardware/software requirements will provide the customers satisfaction with their visit to the hospitals.

Keywords— Multi-agent Systems, Scheduling, Optimization

1. Introduction

In hospitals the use of the various services in the hospitals like blood testing, scans, X-Rays etc. can be optimized if scheduled properly. Patient Scheduling is the process of scheduling the patients by creating an optimal schedule for the patients in using the resources in hospitals. The real time scheduling with various constraints like emergency situations like accident case makes this problem complex. More generally, scheduling problems are subjected to certain constraints to optimize some objective function [1]. This is usually done manually in most systems to load balance a system effectively. Effective Patient Scheduling can improve hospital service quality and operation efficiency. The manually managed and traditional scheduling algorithms needs high coordination, time and manpower to achieve its efficiency. In the autonomous system designed by [6] it's proved that it have improved the service delivery by generating and maintaining proper schedule in spite various scheduling problem differ in its characteristics. The finding out of MAS has derived the way of deploying software agent to solve the scheduling problem [2]. Agents are a natural choice because they can be used to represent a distributed environment well and they have the ability in coordinating among the other agents in dynamic environment. Agents based model follows Belief-Desires-Intentions (BDI) architecture that supports in developing intelligence to the autonomous systems. This helps in selecting and achieving plans without disturbing the real time systems execution [3]. The adapting and dynamic property of the autonomous systems made these agents meet out the real time needs [4].

The architectural components like Belief that maintains the internal state of the system, Desires that represents motivational of the desires state of the agent and Intentions that helps to deliberate the state of the goal make this autonomous system adaptable and fault tolerant [5].Group of heterogeneous agents form an Multi Agent Systems(MAS). MAS allows the sub-problems of a constraint satisfaction problem to be subcontracted to different problem-solving agents with their own interests and goals. The most important reason to use MAS when designing a system is that some domains require it people who have various BDI but intent towards achieving a common goal without conflict. These systems can be well developed using the BDI architecture of MAS. The software agents can be programmed to act with their own capabilities and abilities. Multi-agent systems also have more scalability when agents have to be added. Systems whose capabilities and parameters are likely to need to change over time or across agents can also benefit from this advantage of MAC [6]. This paper describes about the basic of scheduling in chapter 1 follow with the survey of scheduling in healthcare domain and system design for optimization for patient scheduling using multi agents ended with its implementation.

2.Background Studies

Effective resource utilization can improve hospital service quality and operation efficiency. Traditional algorithms like FIFO, linear programming are been used manually and computerized used to solve this in many hospitals. The existing studies try to not only improve the service quality but also to help the allocators do their works efficiently. However, each problem has its own characteristics[7]. Beside those different application's sectors, these approaches are linked by a common objective, which is also ours: create cooperation between agents in systems where decisions and information are highly distributed .The auction based system can help in deriving the best optimized solution for any real time scenario[8]. Using a specific agent as coordinator could be considered for preparing the plan and different care providers' supply numerous services [9].The negotiation is highlighted by the agents to compromise the policies between the doctors and the various agents which is made for administrators[10]. Objectives used for the proposed systems to reduce the patient waiting time while using the labs is mentioned in [11]. However, the specific problem characteristics including time and hardware/software requirements for generating an effective and efficient schedule that are seldom addressed in the existing studies. Recent research has taken a new dimension on the possibility of deploying software agents to solve the allocation problem [12] in the various domain. Agents are a natural choice because they are known to represent a distributed environment well and they are also capable of coordinating with each other which works well in dynamic environment[13][14]. Agents also has the capability to interact with each and coordinate to achieve a common goal with intelligence adds its social behaviour and dynamic nature[15].

Predictive scheduling techniques are used for static environments, but rescheduling is often done to aid the recovery of system from disruptions. Reactive scheduling in dynamic environments often uses a variety of priority rules. R. Haupt [16] has given a list of 26 different scheduling rules. Some of the popular dispatching rules are FCFS (First Come First Serve), MS (Minimum Slack), SPT (shortest processing time), LPT (longest processing time), WPT (Weighted Processing Time), EDD (earliest due date), SL (smallest slack) and CR (critical ratio). The advantage of traditional scheduling algorithm is that these algorithms can be guaranteed to produce a schedule which is robust and reliable. However, many of these algorithms require a central computational unit and do not represent the distributed nature of the patient scheduling problem. Also they are inflexible, expensive and they do not handle disruptions very well.

3. System Description

From the earlier discussion we can conclude that although the hospitals are trying to provide better services to the patients, but they are still not able to provide the required services in optimized manner. Motivated by these factors, my research aim is use multiagent coordination and optimization in healthcare domain. The main objective of the proposed research is to develop a computerized system to achieve the following:

- Patient satisfaction by providing the required resources at the right time.
- Reducing the waiting time of the patient
- Automatic negotiation between the agents avoid human conflicts during the criticality of each and every patient
- Increase the usage of resources by providing all time availability.
- Prioritizing the usage pattern

In the proposed system each resource and patient is represented as agents, which has the ability to coordinate among them. The optimal schedule is calculated in minimal time using the genetic algorithm (GA), an efficient search-based optimization technique. When a patient arrives, he/she is served by a common agent where the priority, resource requirement, processing time, and sequence are assigned to each patient. An optimal schedule is generated using genetic algorithm. The generated schedule is compared with the exact methods like dynamic integer programming for comparative study of time complexity. This helps to prove the effectiveness of the algorithm. The framework in which the whole system works is given in the fig. 3.1. When a patient arrives, he/she is served by a common agent (CA), a general physician who asks the patient about their overall health condition. After the consultation with the general physician, the CA makes the entry about the various operations they have to undergo for a specific time in different resources like scan and blood test.

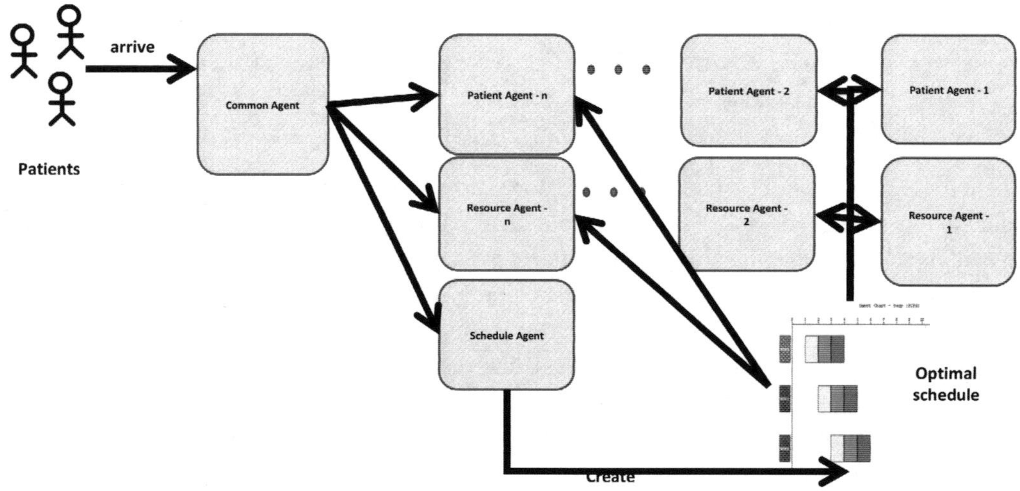

Figure 3.1: System Architecture

Thus, the overall plan of patient's treatment is generated by the CA. The CA also creates individual agents for each patient called patient agent (PA) and resource agent (RA) for all available resources in the hospital domain. PA interacts with the scheduling agent (SA) supporting the distributed nature of the problem. The treatment plan generated by CA is communicated to SA, where genetic algorithm is used to find the optimal schedule at that given time. The generated schedule is then informed to the PA and RA for their usage. The priority of the patients and resources are also allocated by the CA.

Genetic Algorithm (GA) is one of the methods that demonstrated significant improvement in various related domains like job shop scheduling. The initial path was reported by Davis who proposed a new encoding scheme for scheduling problem into the GA solution space and achieved an effective time based schedule without violating the constraint[17]. On the research towards identifying an optimal schedule in dynamic scheduling problems using GA various contributions were made in the areas like chromosome representation, cross over operators, mutation operators and hybrid GA operators. Powered with a minimal time and hardware/software requirement, GA can find its place for generating optimal schedules in health care domains. Making the schedule distributed through the use of multi-agents i.e. accessible will reduce the manual works like telephone operator, crowding at a place and minimizes the patient waiting time and increase the staff productivity in the hospitals [4].

The scheduling agent generates an optimal schedule by getting the treatment plan for each patient from the common agent in minimal time using GA. This works by defining the goal in the form of a quality criteria and move forward by a stepwise refinement to reach a better solution. The generalized algorithm for the GA based optimization proposed for patient scheduling is as follows shown in Figure 3.1.

```
Read Parameters
BEGIN
Generate initial population of size 50 by random from possible permutations
Compute the fitness of each schedule
WHILE NOT finished DO
BEGIN
FOR population_size/2 DO
BEGIN (if crossover rate is less than 0.6 to 1.0 )
Select 2 individuals for crossover
Combine the two individuals by order based crossover
Compute the fitness of offspring generated
Insert offspring in new generation
END
FOR each best individual apply Mutation
BEGIN (if mutation rate is less than 0.001 )
Select individual with high fitness value
Apply insertion mutation
Compute Fitness
Insert Offspring in new generation
END
IF population has converged THEN
finished := TRUE
END
END
Output the best schedule with high fitness
```

Figure 3.2: Algorithm for Schedule Generation

The generation of GA based schedules is robust and can be successful with a wide range of problem areas, including those which are difficult for other methods to solve. Initially whole new populations of possible schedules are produced. The best individuals with high fitness were selected and made to reproduce new generation of off springs. The improvement process is accomplished by GA operators such as Crossover and Mutation. The converged optimal schedule is compared for its efficiency with schedules generated by traditional scheduling algorithms like First-Come-First-Serve and Earliest-deadline-first for time complexity. The exploration and exploitation used by GA makes this algorithm a good solution to calculate an optimal schedule for patient scheduling.

4. Implementation

A GA based Optimization for patient scheduling using multi-agents is designed using Java Agent Development Environment (JADE). Each resource and patient is represented as agents, has the ability to coordinate among them. The optimal schedule is generated in minimal time using the genetic algorithm, an efficient search-based optimization technique. When a patient arrives, each person is served by a Common agent where the priority, resource requirement, processing time, and sequence are assigned to each patient. An optimal schedule is generated using GA by a Schedule agent. The schedule generated is compared with the exact methods like dynamic integer programming and traditional scheduling techniques which includes First-come First-serve and EDF for comparative study of time complexity to prove the advantages of the proposed system. Different genetic encoding is used to generate an optimal solution. Choosing the representation remain critical in the solution. The genes of the chromosome represent the completion time of the operation in each patients.

The system is constructed using the agent framework JADE in windows operating system. An agent is created in JADE by defining a class that extends jade.core.Agent and implements the setup( ) method as shown below. A sample ACL.INFORM message is sent by a common agent to the scheduling agent is given below.

```
public class MyAgent extends Agent
{
protected void setup()
{
System.out.println("Adding waker behaviour");
addBehaviour(new WakerBehaviour(this, 10000)
        {
        protected void handleElapsedTimeout()
        {
        perform operation X
        }} );
}}
ACLMessage msg = new ACLMessage(ACLMessage.INFORM);
msg.addReceiver(new AID("Peter", AID.ISLOCALNAME));
msg.setLanguage("English");
msg.setOntology("Weather-forecast-ontology");
msg.setContent("Today it's raining");
send(msg);
```

Figure 4.1: Sample of Patient name and number of resource agent

The experimental design of patient scheduling system creates two agents named: Common Agent (CA) and Schedule Agent (SA) under the main container and number of patient agents with patient name and number of resource agent with resource name under two separate containers as shown in figure 4.1. The CA starts the JADE runtime environment. It supports the patient and resource registration by entering the related data into the database. It also creates individual agents for each patient and resource entry. It informs the SA to create the optimal schedule using GA through ACL (Agent Communication Language) messages. The optimized schedule generated by the scheduling agent is communicated to all registered patients and resource agents through ACL.INFORM per formative message.

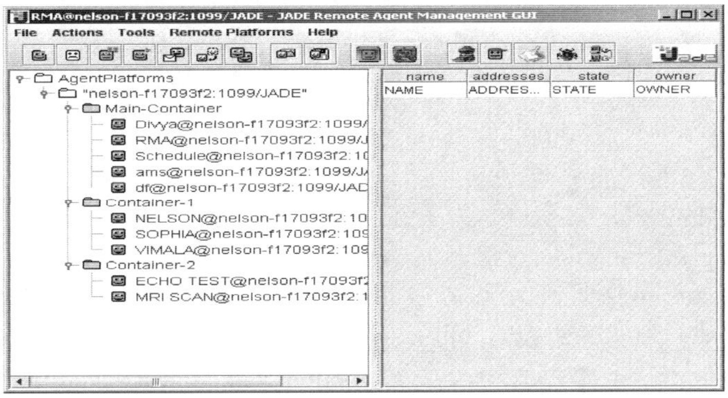

Figure 4.2: Patient scheduling runtime environment in JADE

The Optimal schedule for patient scheduling problem is solved using GA in minimal time. The Scheduling Agent gathers the patient and resource data from database and generates different possible schedules randomly. The schedules are encoded and evaluated using the fitness function. The algorithm will take those schedules and allows them to reproduce by subjecting it to genetic operators like Crossover and Mutation. This goes on until certain criterion is met such as finding a best solution or reaching a fixed number of iterations. The simulation can be run with the experimental inputs. The results obtained should be compared against traditional scheduling algorithms like First-Come-First-Serve and Earliest-deadline-first. The overall completion time is used as a performance metric for the comparative study. LEKIN software can be used to obtain the result for traditional scheduling algorithms. The figure 4.3 shows the screenshot of the experimentation outputs.

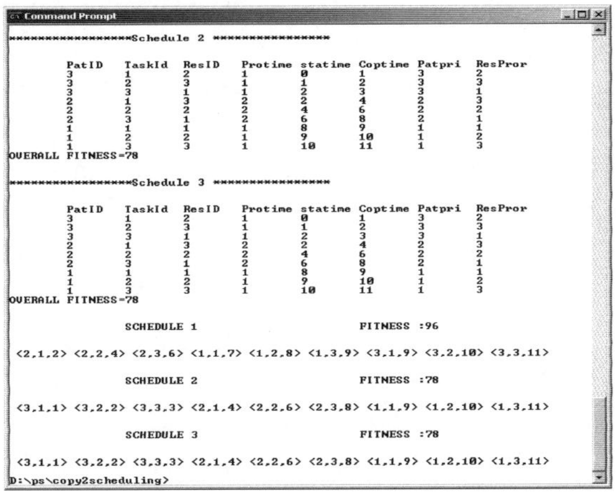

IOP Conf. Series: Materials Science and Engineering 1110 (2021) 012012 doi:10.1088/1757-899X/1110/1/012012

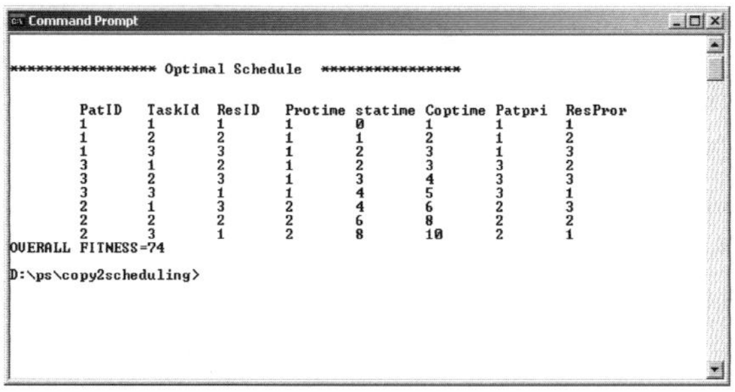

Figure 4.3: Optimal Schedule Generation

5. Conclusion

The multi-agent based solution for patient scheduling has its advantage during the real time situation like scheduling in case of emergency situations. The social and dynamic nature of agents in a multi-agent system makes the task easier. The mobile agent helps in announcing the updated. The schedule to the patients in a timely approach. A proper selection of Genetic algorithm parameters for this application is still an open issue. The crossover and mutation operators presented in this paper should be further investigated in order to improve the performance of the proposed system. Heuristic optimization techniques will surely benefit the time management in constructing an efficient schedule in the very large combinatorial search problems like patient scheduling. Hence in this highly dynamic domain optimal schedule generation with minimal time and hardware/software requirements will improve the quality of livelihood.

References

[1] D. Karger, C. Stein, and J. Wein, "Scheduling Algorithms 1 Introduction," *Environment*, pp. 1–52.

[2] I. Vermeulen, S. Bohte, K. Somefun, and H. La Poutré, "Multi-agent Pareto appointment exchanging in hospital patient scheduling," *Serv. Oriented Comput. Appl.*, vol. 1, no. 3, pp. 185–196, 2007.

[3] J. M. Corchado, D. I. Tapia, and J. Bajo, "A multi-agent architecture for distributed services and applications," *Int. J. Innov. Comput. Inf. Control*, vol. 8, no. 4, pp. 2453–2476, 2012.

[4] M. Hannebauer, "From Formal Work ow Models to Intelligent Agents The Case Study Modeling Work ows Static Aspects," 1998.

[5] W. Shen, S. Member, L. Wang, and Q. Hao, "International membranes directory," *Filtr. Sep.*, vol. 35, no. 1, pp. 45–58, 1998.

[6] J. Sauer, "Knowledge-Based Systems Techniques and Applications in Scheduling," *Knowledge-Based Syst.*, vol. 231, no. 4741, p. 1554, 2000.

[7] N. Benhajji, D. Roy, and D. Anciaux, "Patient-centered multi agent system for health care," *IFAC-PapersOnLine*, vol. 28, no. 3, pp. 710–714, 2015.

[8] M. Reid, "Multi-Agent System Applications in Healthcare: Current Technology and Future Roadmap," *Procedia Comput. Sci.*, vol. 52, pp. 252–261, Jan. 2015.

[9] H. K. Lee, S. Member, J. Li, A. J. Musa, P. A. Bain, and K. Nelson, "Modeling and Analysis of Patient Transitions in Community Hospitals : A Systems Approach," pp. 1–14, 2017.

[10] D. L. Hudson and M. E. Cohen, "Intelligent agents in home healthcare," *Ann. des Telecommun. Telecommun.*, vol. 65, no. 9–10, pp. 593–600, 2010.

[11] R. J. Vwhpv, C. G. Toader, and S. Busnatu, "Patient Flow Control Using Multi-Agent Systems."

[12] J. Xie and C.-C. Liu, "Multi-agent systems and their applications," *J. Int. Counc. Electr. Eng.*, vol. 7, no. 1, pp. 188–197, Jan. 2017.

[13] D. Borrajo and S. Fernández, "Efficient approaches for multi-agent planning," *Knowl. Inf. Syst.*, vol. 58, no. 2, pp. 425–479, 2019.

[14] J. Terán, J. Aguilar, and M. Cerrada, "Integration in industrial automation based on multi-agent systems using cultural algorithms for optimizing the coordination mechanisms," *Comput. Ind.*, vol. 91, pp. 11–23, 2017.

[15] Y. Lyu et al., "Designing and optimizing a healthcare kiosk for the community," Appl. Ergon., vol. 47, no. 277, pp. 157–169, 2015.

[16] R. Haupt, "A survey of priority rule-based scheduling," OR Spectrum, vol.11, no. 1, Mar. 1989.

[17] P. E. Plsek and T. Greenhalgh, "Complexity science: The challenge of complexity in health care. [Article 1 in series of 4]," BMJ, vol. 323, no. 7313, pp. 625–8, 2001.

Chaotic key generators for data security

BABY H T[1] **SUJATHA B R**[2]

[1]Associate Professor, Dept. of E&CE, GEC, Hassan, Karnataka, India

[2] Professor, Dept. of E&CE, MCE, Hassan, Karnataka, India

* Corresponding author's **Email:** babygowda@gmail.com

Abstract: The cryptography approach plays a very major role in securing systems by providing authentication, integrity, and confidentiality for the data. In recent decades, the cryptographic systems have provided novel and efficient ways of encryption. Most of these cryptographic systems need unpredictable entities-Keys. Hence Random number generator is an essential component for cryptography. In cryptography, generating random numbers is an important task, not only for generating cryptographic keys, but also for implementing cryptographic algorithms and protocols. This paper presents chaotic analog circuits to generate random numbers that is used as keys in the conventional cipher to protect the image data.

Keywords: Cryptography, Random number generator, Encryption, Decryption, Chaotic circuits

1. Introduction

In public and private networks as data traffic rate is increasing, it is imperative to secure the confidentiality of the information saved on and exchanged between data processing units. Cryptography is one technique which provides security for the data. In recent decades, cryptographic systems have suggested so many new and competent secure methods of encryption. In all these methods the generation of a cryptographic key is extremely important and to ensure security these keys should be highly random i.e., essentially it is a string of random bits. This can only be achieved by generating highly random numbers. As such**,** random numbers are the foremost requirement for digital signatures, cryptographic algorithms, and communication network protocols. The good random number generator (RNG) should produce numbers that are statistically random and in-deterministic. The chaos theory defines the in-deterministic behavior of the nonlinear dynamic systems which are sensitive to the initial values with specific conditions [1-4]. The chaotic encryption methods are fast, robust, easy to implement and provides high security for the data. The 1D and 2D logistic standard maps are some of the symmetric cipher method which employs substitution-diffusion method. These algorithms include the initial parameters and number of iterations of the chaotic maps and system parameters. But the encrypted cipher text is not fully secure. Only hardware RNG can meet all of these requirements. Therefore, to have a high degree of information protection, a high-quality hardware-based random number generator is absolutely necessary [5-6]. The hardware-based Random Number

Generator strengthens the applications like e-Business, Web browsing and remote access, which currently uses software-based random number generators. The remaining of the paper is organized as follows. In section-2 related works are discussed, section-3 and 4 simulation results of the circuits and randomness test using NIST tool are discussed. In the Section-5 Encryption process and Decryption of the image using Chaos keys are discussed. The section-6 concludes the overall work.

2. Random number generators

Chaotic based circuits have been designed to generate random numbers. The general block diagram of the chaos based RNG system is shown in Fig 2.1, in which output of a chaotic circuit is sampled to obtain digital signal. Then, a binary bit sequence will be extracted by performing the post-processing methods.

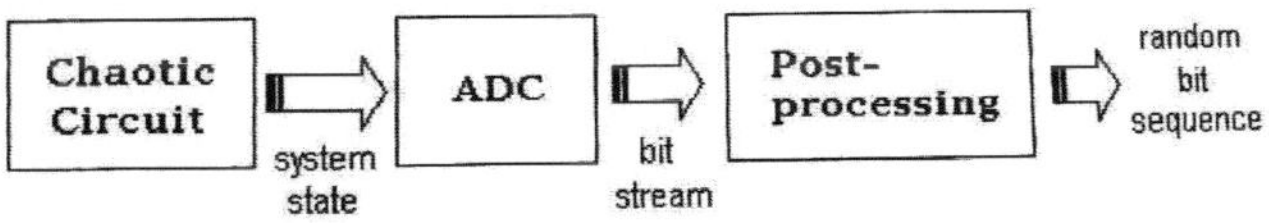

Fig. 2. 1: Chaos-based RNG system

2.1 Colpitts Oscillator as a Chaotic Circuit

An oscillator is one of the electronic circuits which exhibit a rich dynamical behavior like many other third-order oscillator configurations available in the literature. In particular, there is a wide-range of statistical and experimental evidence of continuous chaotic behavior for the Colpitts oscillator that can be used in protected communication systems. This system is generic, non-symmetric and possesses an intrinsic non-linearity which can be represented as the exponential characteristic of the active device [9-10].

The Colpitts oscillator shown in Fig 2.2 comprises bipolar junction transistor (BJT) amplifier with common-base configuration. The resonator circuit includes two capacitances (C_1 and C_2) and an inductance (L), and is mainly used to derive the frequency of oscillations. A portion of the voltage of the LC circuit is feed back to the amplifier. The voltage V_{cc} provides bias to the circuit and the current source I_0 is characterized by a Norton-equivalent conductance G_0. The Colpitts oscillator is normally designed to produce periodic oscillations. However, the oscillator can deliver chaotic waveforms with specific settings of the circuit constraints and appropriate circuitry modifications.

IOP Conf. Series: Materials Science and Engineering 1110 (2021) 012013 doi:10.1088/1757-899X/1110/1/012013

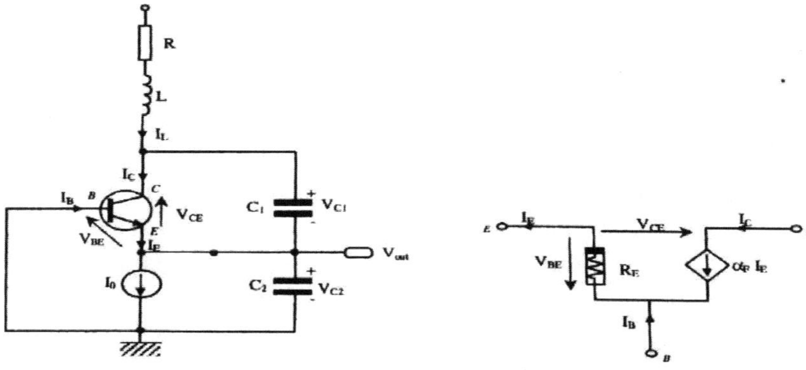

Fig. 2.2: Colpitts oscillator: (a) Circuit schematics (b) B-C configuration

2.2 Amplifier Based Chaotic Autonomous Electronic circuit

This is a simple nonlinear chaotic electronic circuit which comprises Single Amplifier Bi-quad (SAB) which is converted into chaotic oscillator using appropriate passive non-linear element in the form an inductor as shown in Fig 2.3. It is basically an active Band Pass Filter (BPF), which is revised using a passive non-linear p-n junction diode to generate chaotic oscillations, and an inductor as a storage element. The circuit is precisely exhibited using a couple of first-order non-linear differential equations. The circuit shows complex chaotic behavior for a certain set of limit values.

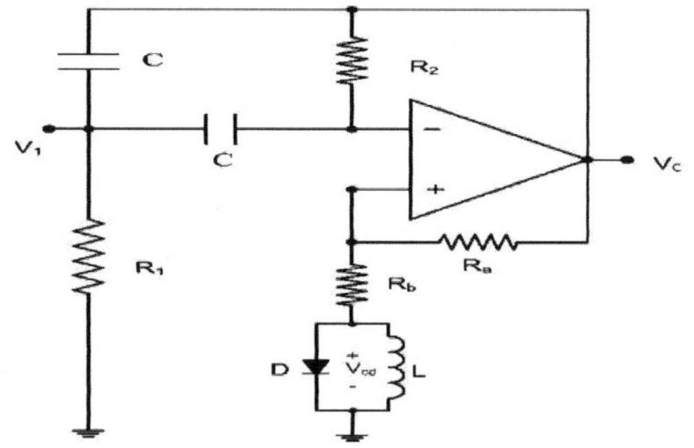

Fig. 2.3: Autonomous chaotic oscillator

The Autonomous chaotic oscillator has (i) A SAB based active BPF comprises of an Op-amp, four resistors (R_a, R_b, R_1, R_2) and two capacitors. (ii) A shunt connection of a P-N junction diode (D) and inductor (L). The diode works as a switch depending on the induced inductor voltage V_{CD} which exists due to the transistance capacitance C_D of the diode. The voltage and current of the autonomous chaotic oscillator are described in equations 1 and 2 respectively.

$$C\frac{dV_0}{dt} = -\frac{2}{R_2}V_0 + \frac{k+1}{R_1}V_1 - \frac{k+1}{R_1}V_{CD}$$

$$\frac{dV_1}{dt} = -\frac{(2k+1)}{(k+1)R_2}V_0 + \frac{k}{R_1}V_1 - \frac{k}{R_1}V_{CD}$$

$$\frac{dI_L}{dt} = \frac{V_{CD}}{L}$$

$$C_D\frac{dV_{CD}}{dt} = \frac{1}{R_1}V_1 - I_L - \frac{1}{R_2}V_{CD} - I_D \quad (1)$$

Here k $=R_b/R_a$. The diode current I_D can be represented as a piece-wise function of linear voltage is given by

$$I_D = \frac{1}{R_D}(V_{CD} - V_Y), \text{ if } V_{CD} \geq V_Y$$

$$= 0 \quad \text{if } V_{CD} < V_Y \quad (2)$$

R_D is forward resistance of the diode and V_Y is voltage drop of the forward biased diode. The Multisim circuit simulator has been used to design the circuit. The simulation circuit uses Op-amp- TL02C2 with ±12V power supply and general-purpose diode along with R_b=500Ω, R_a=2kΩ, C=1nF, L=7.8 mH, R_2=3.1kΩ and R_1=100Ω.

3. Simulation and discussion

Multisim has been used for circuit design of oscillator circuits to generate the random numbers. Python language has been used for checking the randomness of the generated random numbers using standard NIST (National Institute of Standards and Technology) test suite and Dieharder battery test. MATLAB has been used for performing encryption process and decryption of the image data. Fig. 3.1 shows the simulated circuit diagram of colpitt's oscillator that circuit consists Colpitts oscillator, Analog to digital converter (ADC) and Logic analyser. The output of the colpitt's oscillator is shown in Fig 3.2 has converted into discrete signal using ADC and Logic Analyzer has been used to extract the binary bits from the discrete signal.

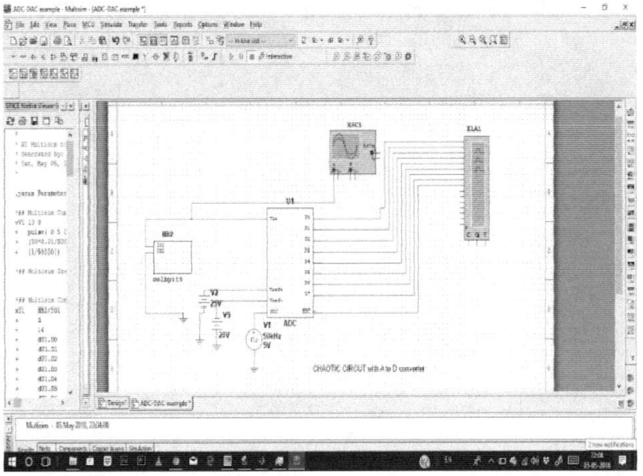

Fig. 3.1: Simulated Circuit diagram of the Random number generator

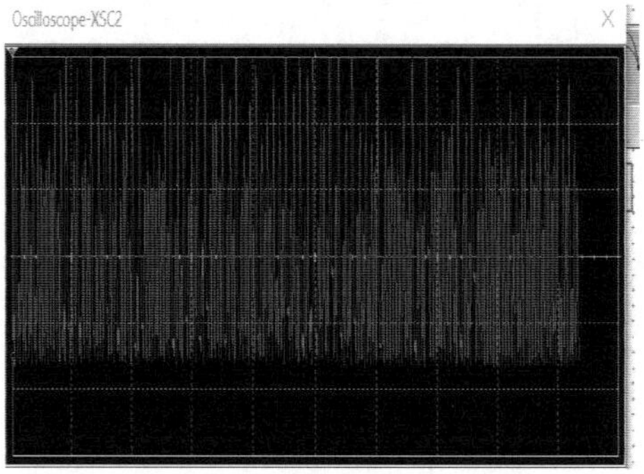

Fig. 3.2: Output waveform of the Colpitts oscillator

Fig. 3.3 shows the simulated circuit of the Amplifier based Autonomous Oscillator and generated waveform. Fig. 3.4 shows the digital waveform from ADC. Binary bits are extracted from the digital waveform using logic analyzer.

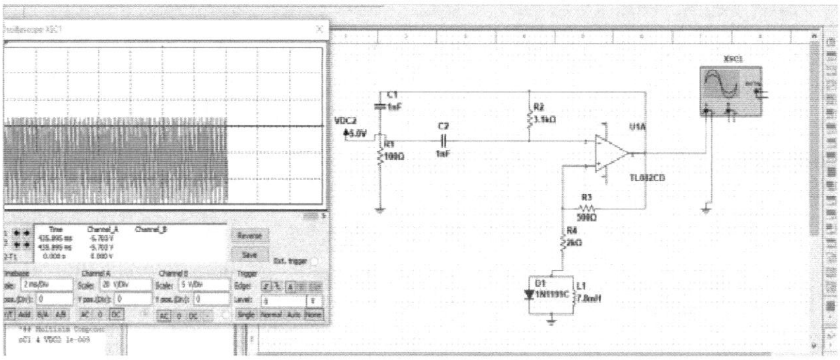

Fig.3.3: Single Amplifier based Autonomous Oscillator output and its waveform

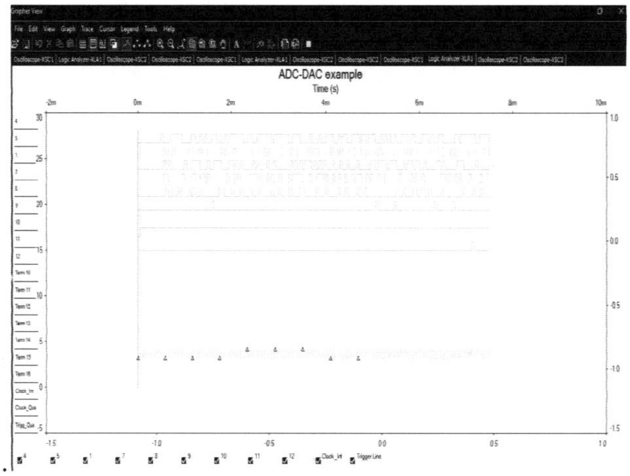

Fig. 3.4: ADC logical waveform

IOP Conf. Series: Materials Science and Engineering 1110 (2021) 012013 doi:10.1088/1757-899X/1110/1/012013

4. Randomness Test

A standard NIST statistical tool and dieharder battery tests are used to analyze the randomness of the binary stream produced by hardware circuits. NIST test focusses on 14 different types of randomness that exists in a sequence and provides percentage of randomness present in it. The 14 tests conducted are described in the table1.

Table 1: Characteristics of the NIST Statistical Tests.

Statistical Test	Characteristics
1. Frequency (Monobit)	Approximately equal number of zeroes and ones
2. Cumulative Sums	Equal distribution of zeroes or ones at the beginning of the sequence
3. Longest Runs of Ones	Equal distribution of long runs of ones or zeroes.
4. Runs	Oscillation between zeroes or ones is too fast or too slow.
5. Matrix Rank	To check linear dependence among fixed length substrings
6. Spectral (Discrete Fourier Transform)	Periodic features in the bit stream.
7. Non-overlapping Template Matchings	Too many occurrences of non-periodic templates.
8. Overlapping Template Matchings	Too many occurrences of m-bit run of ones.
9. Universal Statistical	Compressibility (regularity).
10. Random Excursions	Deviation from the distribution of the number of visits to a certain state
11. Random Excursion Variant	Deviation from the distribution of the total number of visits (across many random walks) to a certain state.
12. Approximate Entropy	Uniform distribution of m-length words.
13. Serial	Non-uniform distribution of m-length words. Similar to Approximate Entropy.
14. Linear Complexity	Deviation from the distribution of the linear complexity for finite length (sub)strings.

Table 2 shows the random numbers extracted from the ADC Circuit. Fig 4.1 (a) and 4.1 (b) shows the NIST tests result and dieharder battery test results for the generated numbers which is computed using Python language. The p value should be greater than 0.01 and less than 1 for randomness. The p values in the results shows that the generated numbers passed all the tests and numbers are highly random in nature. For ex: for the Monobit test P value is 0.4518 and NIST result indicates test is 'PASS', i.e., values generated are random.

Table 2: Generated binary numbers

Sl. No.	Random numbers	Sl. No.	Random numbers	Sl. No.	Random numbers
1	11100011	16	01100011	31	11000011
2	01011011	17	01100011	32	11000011
3	01011011	18	11000011	33	10110011
4	10000011	19	10010011	34	00110011
5	10000011	20	00110011	35	00100011
6	11000011	21	00101011	36	10100011
7	11110011	22	10001011	37	10001011
8	10110011	23	10010111	38	10011011
9	10101011	24	01010111	39	11010011
10	00101011	25	01010011	40	11000011
11	01000011	26	11001011	41	00000011
12	01000011	27	10001011	42	00100011
13	00000011	28	0 0100011	43	01101011
14	10011011	29	01110011	44	01101011
15	11011011	30	01110011		

```
P=0.07
P=0.07
P=0.07
P=0.07
P=0.07
P=0.07
P=0.07

SUMMARY
.......
monobit_test                                 0.4518      PASS
frequency_within_block_test                  0.71523     PASS
runs_test                                    0.7815      PASS
longest_run_ones_in_a_block_test             0.091254    PASS
binary_matrix_rank_test                      0.302412    PASS
dft_test                                     0.152347    PASS
non_overlapping_template_matching_test       0.002345    PASS
overlapping_template_matching_test           0.0         PASS
serial_test                                  0.1354      PASS
approximate_entropy_test                     0.9536      PASS
cumulative_sums_test                         0.531278    PASS
random_excursion_test                        0.017472    PASS
random_excursion_variant_test                0.07        PASS
shree@ sp800_22_tests-master $
```

test_name	ntup	tsamples	psamples	p-value	Assessment
diehard_birthdays	0	100	100	0.70454454	PASSED
diehard_operm5	0	1000000	100	0.37696760	PASSED
diehard_rank_32x32	0	40000	100	0.69136881	PASSED
diehard_rank_6x8	0	100000	100	0.29307364	PASSED
diehard_bitstream	0	2097152	100	0.78273110	PASSED
diehard_opso	0	2097152	100	0.95971692	PASSED
diehard_oqso	0	2097152	100	0.98289579	PASSED
diehard_dna	0	2097152	100	0.28380649	PASSED
diehard_count_1s_str	0	256000	100	0.88955268	PASSED
diehard_count_1s_byt	0	256000	100	0.18091292	PASSED
diehard_parking_lot	0	12000	100	0.16687506	PASSED
diehard_2dsphere	2	8000	100	0.18822358	PASSED
diehard_3dsphere	3	4000	100	0.65369047	PASSED
diehard_squeeze	0	100000	100	0.92414313	PASSED
diehard_sums	0	100	100	0.52629376	PASSED
diehard_runs	0	100000	100	0.38910629	PASSED
diehard_craps	0	200000	100	0.55714545	PASSED
marsaglia_tsang_gcd	0	10000000	100	0.47010128	PASSED
marsaglia_tsang_gcd	0	10000000	100	0.82912277	PASSED
sts_monobit	1	100000	100	0.47839627	PASSED
sts_runs	2	100000	100	0.90978888	PASSED
sts_serial	1	100000	100	0.96300217	PASSED
rgb_bitdist	1	100000	100	0.38562642	PASSED
rgb_minimum_distance	2	10000	1000	0.58870989	PASSED
rgb_permutations	2	100000	100	0.76706057	PASSED
rgb_lagged_sum	0	1000000	100	0.45608115	PASSED
rgb_kstest_test	0	10000	1000	0.12803282	PASSED
dab_bytedistrib	0	51200000	1	0.92437574	PASSED
dab_dct	256	50000	1	0.13841380	PASSED
dab_filltree	32	15000000	1	0.76633250	PASSED
dab_monobit2	12	65000000	1	0.50958303	PASSED

Fig. 4.1: a) NIST test results b) Dieharder battery test results

5. Encryption and Decryption using chaos keys

The lena image shown in Fig 5.1(a) is considered as an input image**.** By performing XOR operations, the input image is encrypted using the generated random numbers as keys and same has been used to perform the decryption process. Fig. 5.1(b) and Fig. 5.1(c) shows the encrypted and decrypted images respectively. There is a great resemblance between the input image and Decrypted image which represents the input image is recovered correctly.

Fig. 5.1: a) Input image b) Encrypted image c) Decrypted image

The Number of pixel change rate (NPCR) is one of the parameters used to analyze the effect of changing one pixel in both plain and cipher images. The NPCR represents the number of pixels differs between the two ciphered images obtained from one-bit differed input image. If C1 and C2 are the two different ciphered images, then the NPCR will be given by,

$$NPCR = \sum_{k=1}^{M} \sum_{l=1}^{N} \frac{d(k,l)}{MXN} \times 100 \qquad (1)$$

d(k,l)= 1 if C1(k, l) = C2(k,l)

= 0 if C1(k,l) ≠ C2(k,l) (2)

The NPCR result obtained is average of 98%, shows that encryption method is very sensitive to minor changes in pixel of the input image.

6. Conclusion

This paper presents chaotic colpitt's oscillator and SAB based chaotic autonomous oscillator circuits to generate binary sequences. The NIST standard tool and Dieharder battery test suite are used to verify the randomness of the generated binary sequences. The results have shown that the binary sequences generated are random in nature and can be used as keys for cryptographic algorithms to secure the data. In this work, image encryption is done using the generated keys and shown that it provides security for the image data. Also, these random sequences can be used for various applications like NONCE generation, One-time pad, password, initialization vectors and parameters of various protocols in crypto systems.

References

[1] Constantin Cehan, Bogdan Cristea, 2002, "Applications of Chaos Theory in cryptography", *International Conference on Military Equipment and Technologies Research*, Romania, pp: 1-5.

[2] Muhammed Rezal, Kamel Ariffin, 2008, "Chaos Based Cryptography an Alternative to Algebraic Cryptography", Research Bulletin Institute for Mathematical Research, Vol **2**, pp: 41-47.

[3] Dubrova E, Teslenko M and Tenhunen H, 2008, "On Analysis and Synthesis of (n, k)-non- linear Feedback Shift Registers," *Design, Test in Europe*, pp: 1286-1291.

[4] Zhu Zhiliang, Chen Dongming, Yang Guangming, 2008, "An improved Image Encryption Algorithm Based on Chaos", *IEEE the 9th International Conference for Young Computer Scientists*, pp: 2792-2796.

[5] Cristian-Iulian, Alexandru, 2009, "Chaos-based Cryptography a Possible Solution for Information security", *Bulletin of the Transilvania*, University of Brasov, vol 2(51), pp: 113-126.

[6] W Jonker and M Petkovic, 2014, "Data Security – Challenges and Research Opportunities", *Springer International Publishing Switzerland*, pp: 9–13.

[7] N N Mosola, M.T Dlamini, Jonathan Blackledge, 2017, "Chaos-based Encryption Keys and Neural Keystore for Cloud-hosted Data Confidentiality", *Telecommunication Networks and Applications Conference (SATNAC, 2017)*, pp.168-173, Southern Africa.

[8] Rim Zahmoul, Ridha Ejbali, Mourad Zaied, 2017, "Image encryption based on new Beta chaotic maps", *Optics and Lasers in Engineering, Elsevier Ltd*, pp: 0143-8166.

[9] Tanmoy Banerjee, Bishnu Charan Sarkar, 2010, "Single amplifier biquad based autonomous electronic oscillators for chaos Generation", *Nonlinear Dynamics*.

[10] Lahcene Merah, Adda Ali-Pacha, Naima Hadj Saidand , Mustafa Mamat, 2013, "A Pseudo Random Number Generator Based on the Chaotic System of Chua's Circuit, and its Real Time FPGA Implementation", *Applied Mathematical Sciences*, HIKARI Ltd, Vol.**7**, No. 55, pp: 2719 – 2734.

Risk Measurement and Cyber security in Industrial Control Systems

[1]Manish Manohar, [2]Srinidhi Hiriyannaiah, [3]Siddesh G M, [4]K. G. Srinivasa
[1,2]Dept. of Computer Science and Engineering Ramaiah Institute of Technology MSR Nagar, Bengaluru, India
[3]Dept. of Information Science & Engineering Ramaiah Institute of Technology MSR Nagar, Bengaluru, India
[4]National Institute of Technical Teachers Training Research, Chandigarh, India
Email:manish.m1138@gmail.com, srinidhi.hiriyannaiah@gmail.com

Abstract. In this new era of expansion in the cyber space, it is not just the IT domain that is vulnerable to cyber-attacks but also the Industrial Control Systems (ICS). ICS is vital in the functioning of almost all the industries either directly or indirectly. As we progress in the direction of Industry 4.0, the challenges with respect to the security of these systems and the data associated with it needs to be addressed. Vulnerability of these systems to various attacks in the past, increases the need for a secured infrastructure. In this paper, the security of ICS along with the importance of software patching and the calculation of the risk associated with the negligence in the process is discussed.

Keywords. Industrial Control Systems (ICS), Vulnerability, Risk assessment, Cyber security, Artificial Intelligence, Software patching

1. Introduction

The term ICS refers to the systems that control the working of the critical infrastructure of the industries. There are various system architectures that could be used to achieve the same. Supervisory management can be achieved by SCADA (Supervisory Control and Data Acquisition) systems. More process specific operations with a rather lower emphasis on the data acquisition can be achieved using DCS (Distributed Control Systems). The Programmable Logic Controllers (PLC) is a control system arrangement that is used in the field level to control the processes and is usually in association with SCADA systems **[1]**. These systems receive values from remote sensors and send control signals to the Final Control Elements (FCE's) such as control valves, according to the programs embedded in it. We are dependent on systems and networks for the functioning of various sectors such as the manufacturing industries, electricity generation plants, sewage treatment plants, chemical industries and various other energy producing plants to simplify the processes involved **[2]**. Some of the attacks that have taken place on Industrial Control Systems are as follows:

Triton attack: In 2017, a Middle Eastern petrochemical facility was attacked by a malware that disabled the safety systems of the devices. This malware became the starting point for the analysis of ICS cyber security in terms of safety systems.**[3]** Disabling the safety systems in ICS is highly dangerous not just for the instruments and the workers in that area but it could also affect a much larger region depending on how critical the control systems are. For example, if the safety systems of a nuclear power plant is disabled, there could be an uncontrolled fission or fusion reaction that could potentially destroy a huge area. It is due to these reasons that the malware is also called as "the world's most murderous malware". **[4]**

Stuxnet Malware: Stuxnet consists of three parts. The first part executes the routines and is called the worm. The second part is a file that executes the copies of the worm and the third part is the rootkit that masks all the files that were created in order to avoid suspicion. This malware is known to target the unpatched Windows systems for automatically executing its processes. The security team usually have a tough time removing this malware due to its intricate design. **[5]**

Duqu Malware: Duqu is an information stealer rootkit. Duqu is a tough bug to detect due to its complex design. This malware was named as Duqu because of the fact that it prefixes the files that is creates with "~DQ". Just like Stuxnet, it exploits the vulnerabilities in the Microsoft Windows systems. Once the dropper file is open in the system, it begins the installation process on its own. The major (and tricky) problem was that it did not start the installation immediately. It began the process only when the computer was left idle for a at least a period of 10 minutes. **[6]** It is important to observe from the critical nature of these malwares that patching of the software in ICS from time to time is very important in order to avoid the high level of damages that would be incurred from the attack.

2.Background

Revolutionary changes have always taken place in the industry whenever new forms of energy or technology was discovered or invented. The growing demand for services/products pushed people to find out innovative methods to complete the tasks in a shorter period of time.
Some of the reasons for the usage of Industrial Control Systems are:

- Reduced labour costs involved in production/maintenance of the infrastructure.
- Time required for the production in assembly lines is less.
- Lower chances of errors as most of the processes are controlled by the machines with high levels of accuracy.

2.1 Intrusion Detection Systems (IDS's)

Intrusion Detection Systems are installed to detect any unwanted traffic entering these systems and send an alert whenever it is detected. There are two main classes of IDS's:

a) Signature analysis IDS: A set of rules are fed into the IDS on the basis of the attacks that are common to the ICS. The IDS can detect only the known attacks. However, unknown forms of attacks may not be detected.

b) Statistical IDS systems: This type of IDS acquires an understanding of the usual functioning of the system by analysing the data and it raises a flag when it detects an attack. This could include both the known or unknown forms of attacks.

The main requirement of the IDS is that it produces really good results and minimal false flags being raised. The chances of false alarms are higher in case of Statistical IDS systems as it may learn some features that may not have been required. **[7]** The idea of using models in determining whether the system is functioning in the way that it is supposed to is not new in the industry. This increases the chances of detecting any illegal intrusions by attackers who try to gain access to these systems by trying different patterns. **[8]** IDS is usually designed to provide a general view of the actions being taken by it in the facility. This may be a problem as it usually fails to detect some of the attacks that target specific components of the ICS. For example, a malware called 'Stuxnet' is intricate in its design and targets particular components of the ICS. It has been reported that it tries to take control of the PLCs. It is important to ensure that there are minimal chances of false data injection attacks on ICS. The malware could continuously change the frequency data from low to high or vice-versa and may therefore pass through the IDS. This could inflict serious damage to the instruments or the people in the surrounding area if the impact is high. **[9]**

2.2 Software Defined Networking firewall for ICS

Dividing the networks into smaller segments on the basis of their purposes in the process is one of the cyber security practices that has been proven to be successful in protecting the ICS. Firewalls are installed in order to filter out the unwanted traffic to the ICS thereby reducing the risk of an attack. Splitting the network into segments helps in preventing the attacks which are based on the scanning of the ICS networks. Network probing attacks are one of the major reasons for the security breaches in these industries according to NCCIC. These attacks are generally performed using network scanning tools such as Nmap, SHODAN and Nessus. **[10]**

2.3 ICS Honeypot Systems

Honeypot systems are those systems that are designed to be vulnerable to attacks and can be scanned by network scanning tools. The main purpose of this type of a system is to direct the attention of the attacker towards the Honeypot system and try various attacks on it. In the process, the security team analyses the types of attacks being used by the attackers. Usually the device to be observed/analysed is cloned and is used as the honeypot system. Even though this is the right method to be followed, it poses a problem in the case of ICS. It is difficult to clone a device in the ICS due to a variety of reasons such as the space that it would require, the costs involved in cloning it as well as the extra maintenance for the honeypot system. An alternative to this could be to use the concept of virtualisation of the honeypots. PLC emulators can be used to achieve the same. This has low costs associated and occupies lesser space. **[11]**

2.4 Artificial Intelligence in ICS

Apart from the traditional methods for preventing cyber-attacks on ICS, there has been a new domain that has emerged. With the increase in computational capacity it is becoming more and more easy to implement AI based solutions for ICS. In the recent years, there has been a surge in the research regarding the utilisation of AI in cyber physical systems. Many researchers are using machine learning models such as Decision trees, SVM, k-NN, random forest, AdaBoost and other deep learning models to detect any anomalies within the system. **[12]** Firewalls are now more stronger due to AI. It can develop its own rules to accept or reject packets entering the network based on the domain name or IP address of the packet. **[13]** However, as the technology advances, attackers have also been trying out new techniques leveraging AI capabilities. One of the examples of an AI based attack is the AI based DDoS (Distributed Denial of Service). DDoS is an attack in which the attacker sends a huge number of requests beyond a point that the server can handle which makes the server to deny service to legitimate users. DDoS is not a new form of attack on cyber physical systems but when it used in conjunction with AI, it is now possible for the program to tweak its parameters on its own without any human intervention making it even more dangerous. It is now easier for the attacker to switch the domain automatically using artificial intelligence with Domain Generation Algorithm (DGA). A static list of domains can be blocked by the defenders but it becomes difficult when there are a huge number of domains which were generated. These generated domains are active for limited periods of time and this time could also be controlled smartly using AI. There are two sides to the use of AI but the advantages of this outweigh the disadvantage and can therefore be considered as a necessary evil. **[14]**

Table 1 gives a comparison between the AI based solutions and the conventional methods for securing the ICS.

Table 1. Comparison of conventional security solutions with AI based solutions for ICS

Conventional security solutions for ICS	AI based solutions for ICS
Capable of detecting anomalies in the system based on the protocols given by the network administrator.	Capable of detecting the anomalies based on the protocols and can detect any new anomalies by learning about the impacts on its own.
Slow in detecting attacks or intrusions.	Faster in detecting attacks or intrusions due to the higher computational capabilities.
Costs incurred is more for securing the systems and maintenance.	Lower cost for securing the systems however, an additional cost for using the cloud or edge-based computing.
There is no need for any huge datasets.	It is important to have good quality datasets to train the model. However, even with the huge amount of data, it may be a problem as the data needs to be structured.
Lower chance for 'False positives' but new attacks may not be blocked.	Higher chances for 'False positives' but there is a better chance that the new attacks will be blocked. It is very important in ICS to make sure that the 'False positives' are minimal as they could lead to a negative economic impact for the industry.

2.5 Managing the risks in ICS

Managing the risks in ICS is important to reduce the impacts of a malware attack by finding the right alternative procedure. One of the ways of measuring the risk in the ICS device is by multiplying the loss with the probability that an attacker would compromise the device. The probability depends on the ease in which the system could be attacked and the skill of the attacker. **[15]** Risk to an ICS includes the probability of converting a source of risk into loss or damage. In order to mitigate the occurrence of a source of risk or hazard in an ICS, various security measures are taken and the control systems are monitored regularly. When the origin of an attack on the ICS is converted to a loss, we can include this under the category of risk. It is very important to take various security measures such as monitoring the control systems regularly and auditing which would help in analysing the source of an attack through the IP address recorded during the process of auditing. The most common way of analysing the risks in ICS is by using the concept of probability. This probability changes from time to time as technology keeps on advancing and the attack strategies change. **[16]** The case study on "Maroochy Water Services, Australia" by MITRE Corporation shows that it is very important to have a no-trust policy within any industry. An insider could cause severe damage to the systems and the environment as ICS are critical systems. **[17]**

3. Proposed methodology

Whenever there is software attached to any device, there is always a chance for unknown backdoors or vulnerabilities to creep into the system. These bugs could prove to be fatal if exploited by attackers. This is especially important when it comes to ICS as these systems have huge workloads. Frequently patching these systems render these instruments unavailable during the entire period of software patching by the security team. **[18]**

Risk Measurement Methodology in Industrial Control Systems (RMMICS):

The risk (r) of the ICS being compromised by attackers is directly proportional to the time taken in days (t) to patch these vulnerabilities:

$$r \propto t \qquad \text{...(1)}$$

Introducing a constant 'c' in order to transform it into an equation:

$$r = c \times t \qquad \text{....(2)}$$

Here, 'c' is the CVSS (Common Vulnerability Scoring System) base score of the vulnerability in the ICS which is a constant for a particular vulnerability. A bug with a high severity has a higher value for 'c' and a bug with a lower severity has a lower value for 'c'. As the value of 'r' increases, there is a higher chance for the ICS being compromised by an attacker. We cannot change the value of 'c' but we can reduce the time taken to patch a vulnerability in the ICS. For example, for c=3.0, the value of 'r' shouldn't cross the 300-400 range (by considering a timeline in the range of 100-115 days) to minimize the risk. This range can vary depending on the history of attacks and the security of the ICS. Table 2 **[19]** shows how the CVSS base score is determined on the basis of the attack vector.

Table 2. Values of 'c' based the severity of the bug.

Value of 'c'	Severity of the bug in the software	Examples
0	None	No bugs
0.1 – 3.9	Low	Language translation error in the software for the controls being used in the ICS
4.0 – 6.9	Medium	Maximum permissible levels of frequency of rotation of a centrifuge is not defined which can be exploited
7.0 – 8.9	High	Low level cryptographic methods used for password authentication
9.0 – 10.0	Critical	Zero day vulnerability that could lead to a root level access of the control systems

Current scenario: Currently ICS related devices are not being patched at the rate that it should be. Many sectors neglect the process as it takes a lot of time and it is a burden for them to patch the systems regularly. Fig 3.1 shows the increase in the risk of the ICS being compromised as the time increases when the CVSS base score is 3.0.

It is important to ensure that the personnel handling the process of patching the software associated with ICS are knowledgeable. In the process of patching, there should be no new vulnerabilities introduced. The system should also be tested before resuming with the activities.

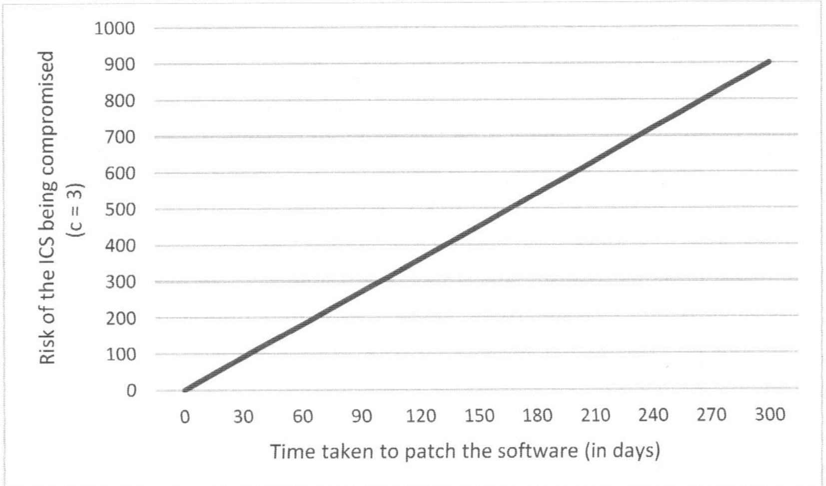

Fig 3.1. Graph for the risk vs time taken to patch the software in ICS (c=3)

4. Experiments and Results

While analysing the various scenarios of attacks that could occur along with the associated CVSS scores, we were able to compile the results in the table 3. The CVSS base score of a vulnerability differs according to the kind of vulnerability as can observed from the table. According to RMMICS, it is advisable to patch the software within the maximum permissible time limit in order to avoid any serious damages to the ICS equipment or the loss of confidential data. The graph corresponding to each of the vulnerabilities shows how the risk increases with time according to the CVSS base score of the vulnerability.

Table 3. Scenarios of attacks along with the risk associated with the attack

Vulnerability in the software	CVSS Base Score (c)	Maximum permissible amount of time to patch the software vulnerability according to **RMMICS**	Graph of risk vs time taken to patch the software vulnerability [X-axis = Time taken to patch the software (in days) Y-axis = Risk of the ICS being compromised (c = CVSS Base Score)]
1) A vulnerability in the ICS called the buffer overflow causes a "denial of service" situation. **[20]** Buffer overflow occurs when a lot of memory is being utilized beyond the permissible limit. By privilege escalation, the attacker could gain admin level access to the system.	7.8 (High)	39 – 52 days	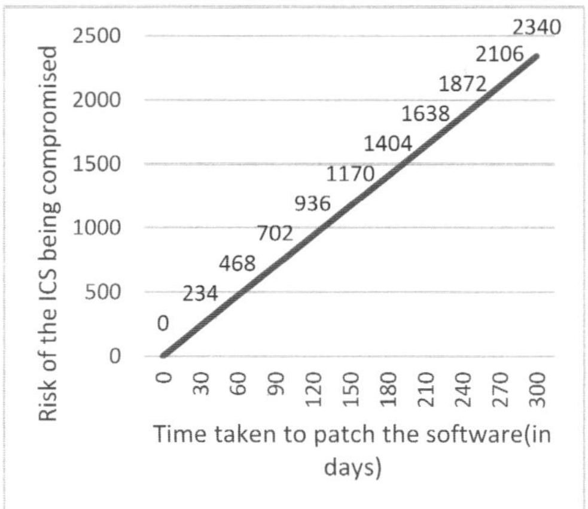

IOP Conf. Series: Materials Science and Engineering 1110 (2021) 012014 doi:10.1088/1757-899X/1110/1/012014

<table>
<tr><td>2) One of the Ethernet/IP adaptor source code was vulnerable to a buffer overflow based on the stack. This vulnerability allows an attacker to send a particular kind of packet that could lead to a Denial of Service. [21]</td><td>9.8 (critical)</td><td>31 – 41 days</td><td>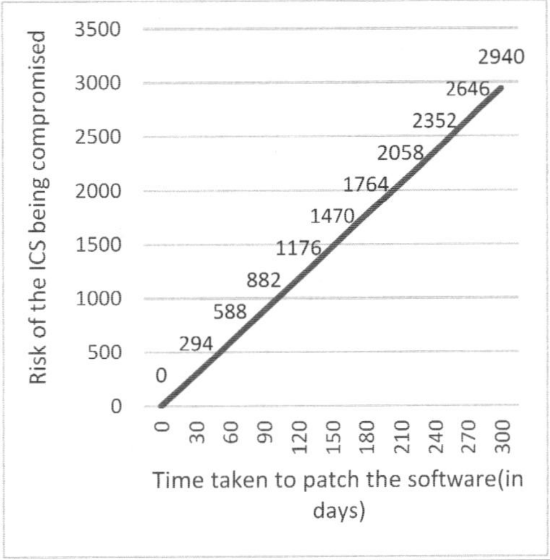
</td></tr>
<tr><td>3) The vulnerability found in Symantec Industrial Control System Protection, versions 6.x.x. The attacker to have higher privileges where the user application or confidential data could be manipulated. [22]</td><td>6.5 (medium)</td><td>47 – 62 days</td><td>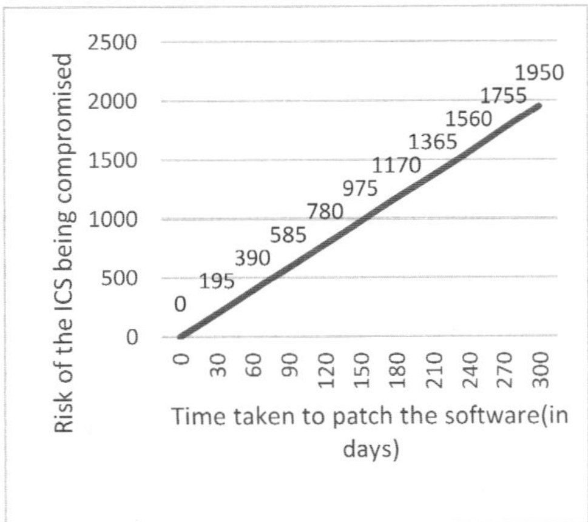
</td></tr>
<tr><td>4) The Martem TELEM GW6 and GWM devices with a particular firmware version and its previous versions allowed for the connection to one or more inputs or outputs and then does not close the connection. This causes a Denial of Service as the inputs and outputs are not available for the next process in the Industrial Control System.[23]</td><td>7.5 (high)</td><td>40 – 54 days</td><td>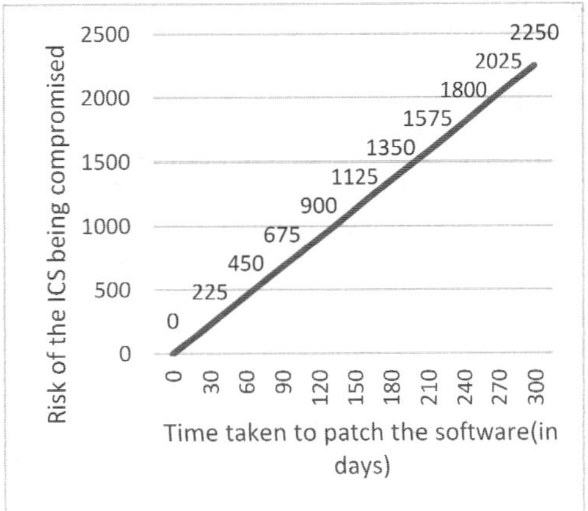
</td></tr>
</table>

5) This vulnerability allowed the attacker who was connected to the Wifi SSID of the IoT device to execute remote code. The attacker could exploit the vulnerability associated with the changing of the name of the Wifi by overflowing this function. Due to this vulnerability, the attacker could access the video feed or pictures (confidential data). **[24]**	8.8 (high)	35 – 46 days	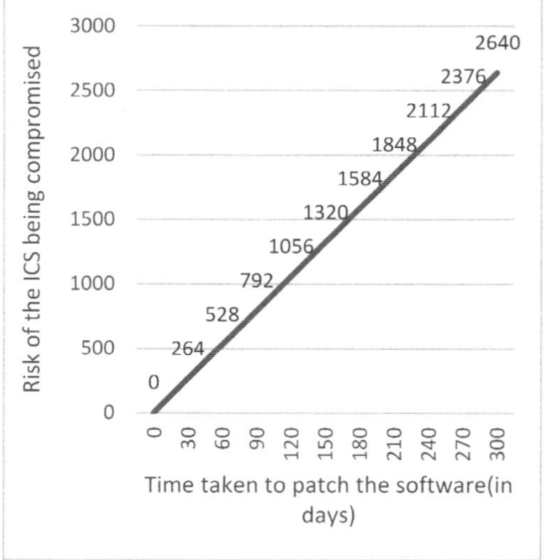

From the table 3, it can be clearly inferred that the value of risk increases withing lesser time period for vulnerabilities that have a high CVSS base score. The maximum permissible amount of time to patch the software vulnerability is calculated approximately by taking a risk value in the range of 300 to 400. The lower limit on the number of days is calculated as follows:

$$\text{Lower limit} = 300/(\text{CVSS base score}) \qquad \ldots(3)$$

The upper limit on the number of days is calculated as follows:

$$\text{Upper limit} = 400/(\text{CVSS base score}) \qquad \ldots(4)$$

This calculation is on the basis of the selected range of permissible risk values of 300-400. However, this range could vary according to the Industrial Control Systems. The permissible risk value could be lower if it is a critical infrastructure and the permissible risk value could be higher if it does not have a huge level of importance.

5. Conclusion

With the increasing dependence on Industrial Control Systems and the various risks/hazards involved in it due to the various attacks, it is extremely important to secure this critical infrastructure. Artificial Intelligence is gaining momentum in the space of ICS. However, it is important to analyse the pros and cons of the type of security solution that would be used. Assessing the risks in case of an attack is essential in taking alternate measures to prevent such an event in future. The experimental results in accordance with RMMICS shows that the risk of an ICS being compromised increases as the time progresses and the CVSS base score of the vulnerability increases. An approximate permissible time limit for patching the software can be obtained by considering a particular range of risk values and the CVSS base score. It is important to get the systems working as soon as possible by patching the bugs in the software regularly and analysing the situation by calculating the risk of the ICS being compromised as well as verifying the security audits in order to minimize the losses.

References

[1] Stouffer, K., Falco, J., & Scarfone, K. (2011). Guide to industrial control systems (ICS) security. NIST special publication, 800(82), 16-16.

[2] Kobara, K. (2016). Cyber physical security for industrial control systems and IoT. IEICE TRANSACTIONS on Information and Systems, 99(4), 787-795.

[3] https://securityboulevard.com/2019/02/oil-gas-cybersecurity-and-process-safety-converge-thanks-to-triton/, accessed June 17th, 2020.

[4] https://www.technologyreview.com/2019/03/05/103328/cybersecurity-critical-infrastructure-triton-malware/ , accessed June 19th, 2020.

[5] https://www.trendmicro.com/vinfo/us/threat-encyclopedia/web-attack/54/stuxnet-malware-targets-scada-systems , accessed June 17th, 2020.

[6] Bencsáth B, Pék G, Buttyán L, Félegyházi M. Duqu: Analysis, detection, and lessons learned. InACM European Workshop on System Security (EuroSec) 2012 Apr 10 (Vol. 2012).

[7] Valdes, A., & Skinner, K. (2000, October). Adaptive, model-based monitoring for cyber attack detection. In International Workshop on Recent Advances in Intrusion Detection (pp. 80-93). Springer, Berlin, Heidelberg.

[8] Cheung S, Dutertre B, Fong M, Lindqvist U, Skinner K, Valdes A. Using model-based intrusion detection for SCADA networks. InProceedings of the SCADA security scientific symposium 2007 Jan (Vol. 46, pp. 1-12).

[9] Wang Y., Xu Z., Zhang J., Xu L., Wang H., Gu G. (2014) SRID: State Relation Based Intrusion Detection for False Data Injection Attacks in SCADA. In: Kutyłowski M., Vaidya J. (eds) Computer Security - ESORICS 2014. ESORICS 2014. Lecture Notes in Computer Science, vol 8713. Springer, Cham

[10] Tsuchiya, A., Fraile, F., Koshijima, I., Ortiz, A., & Poler, R. (2018). Software defined networking firewall for industry 4.0 manufacturing systems. Journal of Industrial Engineering and Management (JIEM), 11(2), 318-333.

[11] Antonioli, D., Agrawal, A., & Tippenhauer, N. O. (2016, October). Towards high-interaction virtual ICS honeypots-in-a-box. In Proceedings of the 2nd ACM Workshop on Cyber-Physical Systems Security and Privacy (pp. 13-22).

[12] Al-Abassi, A., Karimipour, H., Dehghantanha, A., & Parizi, R. M. (2020). An Ensemble Deep Learning-Based Cyber-Attack Detection in Industrial Control System. IEEE Access, 8, 83965-83973.

[13] Chakraborty, Partha & Zahid, Zahidur & Rahman, Saifur. (2019). Building New Generation Firewall Including Artificial Intelligence. International Journal of Computer Applications. 178. 1-7. 10.5120/ijca2019919416.

[14] https://unit42.paloaltonetworks.com/threat-brief-understanding-domain-generation-algorithms-dga/ , accessed June 27th, 2020.

[15] Cárdenas, A. A., Amin, S., Lin, Z. S., Huang, Y. L., Huang, C. Y., & Sastry, S. (2011, March). Attacks against process control systems: risk assessment, detection, and response. In Proceedings of the 6th ACM symposium on information, computer and communications security (pp. 355-366).

[16] Cook, A., Smith, R., Maglaras, L., & Janicke, H. (2016, August). Measuring the risk of cyber-attack in industrial control systems. BCS eWiC.

[17] Abrams, M., & Weiss, J. (2008). Malicious control system cyber security attack case study–Maroochy Water Services, Australia. McLean, VA: The MITRE Corporation.

[18] Wang, B., Li, X., de Aguiar, L. P., Menasche, D. S., & Shafiq, Z. (2017). Characterizing and modeling patching practices of industrial control systems. Proceedings of the ACM on Measurement and Analysis of Computing Systems, 1(1), 1-23.

[19] https://nvd.nist.gov/vuln-metrics/cvss , accessed on Dec 3rd, 2020.
[20] https://nvd.nist.gov/vuln/detail/CVE-2020-7586 , accessed on Dec 1st, 2020.
[21] https://nvd.nist.gov/vuln/detail/CVE-2020-25159 , accessed on Nov 28th, 2020.
[22] https://nvd.nist.gov/vuln/detail/CVE-2019-18380 , accessed on Dec 5th, 2020.
[23] https://nvd.nist.gov/vuln/detail/CVE-2018-10607 , accessed on Dec 4th, 2020.
[24] https://nvd.nist.gov/vuln/detail/CVE-2017-10723 , accessed on Dec 5th, 2020.

IOP Conf. Series: Materials Science and Engineering 1110 (2021) 012015 doi:10.1088/1757-899X/1110/1/012015

A HYBRID CLUSTERING BASED APPROACH TO EXTRACT DRUG ELEMENTS WHICH CAUSES SIDE EFFECTS

[1]Ms. Alpha Vijayan, [2]Dr. Chandrasekar B. S
[1]Research Scholar, Jain Deemed to be University ,[2]Director, Center for Distance Education & Virtual Learning (CDEVL), Jain Deemed to be University

Abstract: Prescription manifestations or drug reactions are an imperative and complex test. Medications are synthetic substances for treating illnesses, however, may actuate unfriendly responses or results. Medication revelation is tedious and works concentrated, and applicant drugs experience the ill effects of likely side. Heaps of affirmed drugs were removed from the market due to sudden results. Since drug results are an incredible worry of the general wellbeing, the recognizable proof of medication results assists with diminishing dangers in drug revelation. With the expansion of medication information, scientists gathered data about endorsed sedates, and recognize possible symptoms of new applicant drugs. Medication disappointments because of unexpected unfriendly impacts at clinical preliminaries present wellbeing hazards for the members and lead to generous monetary misfortunes. Side effect forecast calculations can possibly direct the medication configuration measure. Drug utilized in remedy relies upon the harmony between anticipated favourable circumstances and possible threats.

Keywords: Drug protein interactions, protein- protein interactions, hybrid clustering, structural similarity profiles.

1. Introduction

In the drug business, unrecognized or under revealed ADRs not simply point preventable human suffering and costs to the social protection structure, yet moreover absurdly subvert general society's trust in sedate treatment. Because of genuine ADRs in excess of 2,000,000 hospitalization cases are accounted for consistently [1]. The deadly certified ADRs have transformed into the fourth 6th driving explanations behind death yearly. Studies in Europe and Australia have yielded similar evaluations. It takes various extended lengths of study and security reconnaissance lance to separate these ADRs completely [2]. This delay in understanding squares our ability to perceive, assess, and use ADRs to smooth out medication result assurance and estimation. There is during this way an incredible had the chance to envision and screen a medication's ADRs for the timespan life cycle, from preclinical screening stage to post advertising perception. To lessen ADR-related grimness and mortality, a few computational undertakings to unmistakable verification potential ADRs have been made, including: creating distinctive drug related profiling (e.g., substance profiling, cell response profiling) to envision ADRs at different levels. Likewise, using complex organization induction systems, for instance, coordinate scattering. Atias and Sharan proposed a scattering method in the ADR closeness structure to anticipate score each ADR by expecting that comparable ADRs gets relative scores. Further, distinguishing real signs from suspected Adverse medication occasions and perceiving up-and-comer centers around that have a causal relationship with ADRs.

Powerful ADRs forecast is basic for enhancing patients' human services and quickening the medication advancement process [3]. Diverse computational procedures have been exploited as a part of later past so as to comprehend the system of drug side effect. The information sources used to examine reaction in different studies incorporate chemical information of the drugs and medication targets [4]. The significant reason for side effects of drug is off-target responses. The component of activity of medications is impacted by the genomic heterogeneity of people and affecting compound properties because of modifying smaller scale condition in cell compartments. Thus, reactions "as clinical phenotypes" that emerge in patients can thought to be a sign of complex collaboration of large number of components i.e. genomic highlights, infection state in which drugs are controlled called drug indications, chemical descriptors of medications [5].

2. Literature Review

Bresso et al used an integrative method to manage explain drugs responses. The data was secured from Drugbank and SIDER information base. Bunching of the similar prescriptions is performed by combining the medication targets descriptors and medication fingerprints. Assessment of two AI methodologies ie choice trees and inductive-rationale programming shows that the later outflanked both in execution and to also clarify the valuable relationship in pathways of medication targets and drugs. [6].

Giving scores is an idea related with gaming industry, abused as a portion of this endeavor to elucidate essential linkups between drug illnesses affiliations, prescription and medication result responses for the most part brought about by the meds used for treatment and it can help masters in drug associations to create hypotheses for sedate exposure. An association among pharmacogenomics and responses has been showed up by detaching 244 pharmacogenes which are connected with manifestations of 176 meds from Pharm GKB information base were 28 characteristics are perceived by FDA which are connected with threat of indications [7].

Wei-Po Lee et al. presents the utilization of a cross breed AI way to deal with develop result classifiers utilizing a fitting arrangement of information highlights. It uses the point of view of information examination to research the impact of medication dispersion in the component space, arrange results into a few stretches, receive appropriate procedures for every span, and develop information models in like manner. A progression of tests was led to confirm the relevance of the introduced technique in result forecast. This methodology had the option to consider the qualities of various sorts of results, subsequently accomplish better prescient execution. Also, unique component choice plans were combined with the demonstrating techniques to analyze the relating impacts [8].

Causality examination shows structure learning (CASTLE) device instrument uses both substance and natural properties of meds to choose sub-nuclear atomic pointers of result responses. Figure execution was evaluated on 12 organ-specific ADRs on 830 meds data. The examination pipeline has three phases included extraction, characterization of ADRs utilizing Support vector machines (SVM), advancement investigation was performed for approval and contrasted and OMIM information base outcomes. Notwithstanding the way that the desire execution was promising anyway there was simply most of the way endorsement from improvement examination with OMIM information base represents mandelian legacy qualities in man and contains data identified with mendelian problems and more than 15,000 quality [9].

Song. J [10] depict another registering strategy called PREvalL for the expectation of stores of compound synergists. This methodology was created using an exhaustive game plan of illuminating features in a self-assertive timberland AI plot, removed from different levels, including gathering, structure, and contact association. Wide benchmarking tests eight distinctive datasets relying upon 10-overlay cross-endorsement and autonomous tests, similarly as close to one another display surveys with seven current game plans and structure-based techniques indicated that PREvalL accomplished farsighted execution with land under the curve of the beneficiary working imprint and a zone under the twist of the exactness appraisal. The creators demonstrated that this technique had the decision of catching supportive signs rising up out of various levels, utilizing such differential yet valuable sorts of features and permitting us to improve the introduction of synergistic development forecast totally. The creators perceive this new methodology as a rewarding gadget for both understanding the marvelous gathering structure–working for protein affiliations and advancing portrayal of novel synthetic substances that need utilitarian clarifications.

In [11], in view of the suspicion that comparative medications will in general communicate with comparable proteins and the other way around, they built up a novel computational strategy (specifically MKLC-BiRW) to foresee new medication target collaborations. MKLC-BiRW incorporates different medication related and target-related heterogeneous data source by utilizing the various part learning and grouping strategies to produce the medication and target similitude frameworks, in which the low likeness components are set to zero to assemble the medication and target closeness adjustment organizations. By fusing these medication and target likeness rectification networks with known medication target collaboration bipartite chart, MKLC-BiRW develops the heterogeneous organization on which Bi-arbitrary walk calculation is embraced to construe the potential medication target communications. In the method of building the medication and target closeness mix networks, four medication related wellsprings of the compound structure, drug-illness affiliation, drug-result affiliation and medication ATC code they are utilized to shape the medication comparability coordination framework with the KronRLS-MKL calculation; three objective related wellsprings of the protein groupings, target-sickness affiliation and protein-GO comment they have applied to frame the objective similitude reconciliation lattice. During the time spent changing the closeness coordination lattices to create the likeness amendment grids, the DDI information, known DTI information and PPI information they have used to group the medications and the objectives. Furthermore, the module cohesiveness was received to change the medication and target closeness coordination networks.

Alghamedy [12] wires outfit docking, the use of various protein compliances eliminated from a subatomic components course to perform docking assessments, with extra biomedical data sources and AI figurings to improve the desire for medicine official. The creators found that they can staggeringly assemble the request exactness of a working versus an interruption compound using these procedures over docking scores alone. The best results seen here begin from having an individual protein consistence that produces confining features that partner they with the dynamic versus interruption request, in which case they achieve over 99% exactness. The ability to obviously make exact conjectures taking drugs limiting would consider computational polypharmacological frameworks with pieces of information into response desire, steady repurposing, and medicine feasibility. The creators found in this examination that they can effectively utilize AI to build the forecast of medication authoritative and utilizing drug highlights determined from protein compliances chose from atomic elements directions expands the consistency of the models. Exercises learned in this investigation will be utilized to construct models of various proteins to be utilized in drug disclosure applications. Indeed, they are as of now building up an "all kinase" model to foresee drug repurposing and possible results. The capacity to without a doubt makes accurate desires on drug limitation would consider computational polypharmacological frameworks with pieces of information in response estimating, repurposing sedation, and medication viability. In this examination, they found that they can viably utilize AI to grow the medicine definitive conjecture and use drug features decided from p Exercises learned in this survey will be utilized to fabricate models of various proteins to be utilized in applications for prescription disclosure. The creators are really developing an "all kinase" model to envision repurposing sedation and future indications from here on out.

3. Predicting Side Effects

It is essential to make up an examination pipeline to computationally anticipate drug result indications from different grouped sources. The difficulties saw in the current explores are the nonappearance of prompt genetic information from the patients are not available from open data files. Hence, a significant notion in this examination is those medication results are a gauge of missing inherited information from the patients. The crucial exploration question is that whether the meds and prescription signs are an insightful information for the medication result announced with the drugs. The turn of the examination is to arrange ADRs related with drug signs and substance descriptors. The data sources used to anticipate the side effects are the known as informational collections or medication infection's affiliations and fingerprints/blend descriptors of the drugs. Ten exceptional yet ordinary side effects which were used for this assessment are explicitly Migraine, Unsteadiness, Shortcoming, Stomach Torment, Nausea, Constant Weakness, and Looseness of the digestion tracts, Rashes, Dermatitis, and Spewing. These manifestations with most astonishing change can be picked for this assessment. Unequivocally, data from healing indications of meds close by their creation properties (substance descriptors) are used to foresee clinical aggregates (drug result) of meds.

IOP Conf. Series: Materials Science and Engineering 1110 (2021) 012015 doi:10.1088/1757-899X/1110/1/012015

4. Methodology

In ongoing past, a couple of medicine information bases have been created to empower the assessment that contains advanced drugs and remarked on manifestations or symptoms of the medications. A combination of medicine information can be removed from data sets. The bases of medications are ordinarily estimated as the most imperative factor for sedate responses or results. Medicine targets are commonly connected with a particular metabolic or hailing pathway, and may offer the crucial hint to sedate results. Medicine carriers are imparted in various tissues, and expect indispensable parts in quiet absorption, movement, and release. Meds more regularly than not experience sedates assimilation to be naturally unique, and the synthetic compounds may affect the processing and start responses.

There is assorted prescription data, including drug base data, drug target data, drug target data, quiet protein data, calm pathway data and medicine sign data, which give particular features to portray drugs. By using these boundaries or highlights, the medications utilized can be meant as a component vector, whose estimations show the closeness or nonattendance of relating portions.

Computational strategies hold extraordinary guarantee for moderating the wellbeing and money related dangers of medication advancement by foreseeing conceivable results prior to going into the clinical preliminaries. A few learning-based strategies have been proposed for anticipating the symptoms of medications dependent on different highlights, for example, synthetic structures of medications, drug-protein connections, protein-protein associations, movement in metabolic organizations, pathways, aggregate data and quality comments.

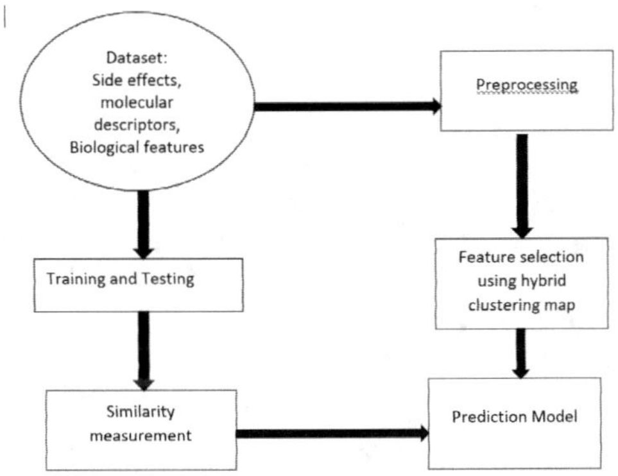

Fig 4.1: Model for extracting side effects

To recognize the names of medications in sentences, official names of medications are collected from DrugBank gives extensive medication data in bioinformatics and cheminformatics.

The issue of result expectation is displayed as a multi-mark classification task. For a given medication i, the objective name is a parallel vector,

$$z_i = [z_{i,1}, z_{i,2}, \ldots z_{i,d}]$$

where d is the quantity of results and $z_{i,j} = 1$ shows that the medication i has result j; $z_{i,j} = 0$ demonstrates in any case. The dataset contains n tests (medicates), each spoke to by a couple of medication include vector x_i and a going with result vector (classes).

Quality images that give data on hereditary variety in drug reactions are gathered. A 3-overlap cross- $z_i : (x_i, z_i)_{i=1}^{n}$ approval is utilized to assess performance. The folds are separated dependent on drugs. All investigations of a solitary medication are either totally in the preparation set or totally in the test set, and accordingly, a model is required to foresee the results of beforehand inconspicuous medications at test time.

A hybrid bunching model is used as a closeness model to distinguish drugs with nearer connections. Unaided bunching is a way to deal with gathering comparative items with no earlier information on their group names. Articles that are in a given group ought to exhibit higher likeness to one another and a moderately higher disparity with the items in different bunches. Table 1 contains some of the drug components and side effects.

Table 1: Drug components with side effects

Drug Component	Side Effects
rosuvastatin	myalgia, abdominal pain, asthenia, nausea
duloxetine	asthenia, Diarrhea, dizziness, Drowsiness, Fatigue, Hypersomnia, Insomnia, Nausea, sedated state, headache, xerostomia, agitation, erectile dysfunction, nervousness, psychomotor agitation, tension, vomiting, abdominal pain, anorexia, decreased appetite, decreased libido, hyperhidrosis, loss of libido, restlessness
escitalopram	diarrhea, drowsiness, headache, insomnia, dizziness, dyspepsia, fatigue, decreased libido, diaphoresis, and xerostomia
atorvastatin	hemorrhagic stroke, arthralgia, diarrhea, and nasopharyngitis, urinary tract infection, insomnia, limb pain, muscle spasm, musculoskeletal pain, myalgia, and nausea
lisinopril	dizziness, hypotension, hyperkalemia, increased blood urea nitrogen, and increased serum creatinine, headache
melatonin	daytime drowsiness, depressed mood, feeling irritable, stomach pain, headache, dizziness.

IOP Conf. Series: Materials Science and Engineering 1110 (2021) 012015 doi:10.1088/1757-899X/1110/1/012015

Table 2: Mapping sample

nct_id	drugbank_id	drugbank_name	doid_code	doid_name
NCT00275262	DB00007	Leuprolide	DOID:0060058	Lymphoma
NCT00003728	DB00023	Asparaginase	DOID:0060058	Lymphoma
NCT00004157	DB00038	Oprelvekin	DOID:0060058	Lymphoma
NCT00002649	DB00041	Aldesleukin	DOID:0060058	Lymphoma
NCT00057811	DB00049	Rasburicase	DOID:0060058	Lymphoma
NCT01476839	DB00074	Basiliximab	DOID:0060058	lymphoma
NCT00038883	DB00087	Alemtuzumab	DOID:0060058	lymphoma
NCT00002759	DB00091	Cyclosporine	DOID:0060058	lymphoma
NCT00002779	DB00104	Octreotide	DOID:0060058	lymphoma
NCT00001249	DB00111	Daclizumab	DOID:0060058	lymphoma
NCT02483000	DB00121	Biotin	DOID:0060058	lymphoma
NCT01983969	DB00147	Pyridoxal	DOID:0060058	lymphoma
NCT02701673	DB00165	Pyridoxine	DOID:0060058	lymphoma
NCT01787409	DB00169	Cholecalciferol	DOID:0060058	Lymphoma
NCT00005799	DB00194	Vidarabine	DOID:0060058	Lymphoma
NCT02532192	DB00196	Fluconazole	DOID:0060058	lymphoma
NCT00364923	DB00200	Hydroxocobalamin	DOID:0060058	lymphoma

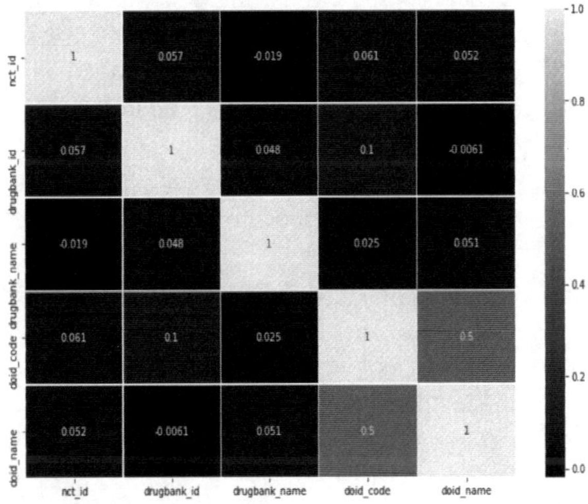

Fig 4.2: Correlation Chart

5. Analysis

The model coordinates the bunching approach with AI calculation and thinks about the accompanying cycle: the qualities are chosen, test sets and preparing sets created, AI calculation determination, planning fitting expectation model, surveying the presentation of the model. It thinks about different results (arrangement marks), and in this manner it is important to consider unmistakable boundaries, for example, precision and effectiveness. In this manner, considering these elements the examination will build up an anticipating model by fusing bunching approach for include choice. AI is coordinated to build up an expectation to choose the ideal measurement subset and create multi-name order model. This is found to adequately look through the intriguing space and tackle complex issues without requiring the former information about the space and the issues.

6.Conclusion

The work plays out a deliberate assessment of different datasets on drug forecast and its results. It builds up a strategy dependent on incorporation of grouping approach for drug result expectation dependent on AI strategies. It will deliver high exactness exhibitions just as the intelligible outcomes that will uncover the causes and results. This methodology is to consolidate different highlights effectively and use them as base indicators. In this manner, the exploration is found to blend base indicators and build up the last expectation models is created for drug result forecast.

References

1. Edwards, I.R. & Aronson, J.K. (2000), Adverse drug reactions: definitions, diagnosis, and management. Lancet356, 1255–1259.
2. Chiang, A.P. & Butte, A.J. (2009). Data-driven methods to discover molecular determinantsof serious adverse drug events.Clin. Pharmacol. Ther.85, 259–268.
3. Leone, R., Sottosanti, L., Iorio, M. L., Santuccio, C., Conforti, A., Sabatini. V,Venegoni, M. (2008). Drug-related deaths. Drug Safety, 31 (8), 703–713.
4. DiMasi, J. A. (2002). The value of improving the productivity of the drug development process. Pharmacoeconomics, 20 (3), 1–10.
5. Schuster, D., Laggner, C., & Langer, T. (2008). Why drugs fail–a study on side effects in new chemical entities. Antitargets. Prediction and Prevention of Drug Side Effects, 3–22.
6. Bresso, E., Grisoni, R., Marchetti, G., Karaboga, A. S., Souchet, M., Devignes, M.-D., & Smaïl-Tabbone, M. (2013). Integrative relational machine-learning for understanding drug side-effect profiles. BMC bioinformatics, 14 (1), 207.
7. Zhou, Z.-W., Chen, X.-W., Sneed, K. B., Yang, Y.-X., Zhang, X., He, Z.-X., Zhou, S.-F. (2015). Clinical association between pharmacogenomics and adverse drug reactions. Drugs, 75 (6), 589–631.
8. Wei Po Lee, Jhih Yuan Huang, Hsuan Hao Chang, King-The Kee, Chao-Ti Lai, (2017). Predicting drug side effects using data analytics and the integration of multastrology
9. e data sources. IEEE Access, 5, 20449-20462.
10. Liu, M., Cai, R., Hu, Y., Matheny, M. E., Sun, J., Hu, J., & Xu, H. (2014). Determining molecular predictors of adverse drug reactions with causality analysis based on structure learning.
11. Song, J., Li, F., Takemoto, K., Haffari, G., Akutsu, T., Chou, K. C., & Webb, G. I. (2018). PREvaIL, an integrative approach for inferring catalytic residues using sequence, structural, and network features in a machine-learning framework. Journal of theoretical biology, 443, 125-137.
12. Yan, X. Y., Zhang, S. W., & He, C. R. (2019). Prediction of drug-target interaction by integrating diverse heterogeneous information source with multiple kernel learning and clustering methods. Computational biology and chemistry, 78, 460-467
13. Alghamedy, F., Bopaiah, J., Jones, D., Zhang, X., Weiss, H. L., & Ellingson, S. R. (2018). Incorporating protein dynamics through ensemble docking in machine learning

IOP Conf. Series: Materials Science and Engineering 1110 (2021) 012015 doi:10.1088/1757-899X/1110/1/012015

models to predict drug binding. AMIA Summits on Translational Science Proceedings , 2018, 26.
14. Chen, H., Engkvist, O., Wang, Y., Olivecrona, M., & Blaschke, T. (2018). The rise of deep learning in drug discovery. Drug discovery today, 23(6), 1241-1250.
15. Liang, Z., Huang, J. X., Zeng, X., & Zhang, G. (2016). Dl-adr: a novel deep learning model for classifying genomic variants into adverse drug reactions. BMC medical genomics, 9 (2), 48.
16. Ching, T., Himmelstein, D. S., Beaulieu-Jones, B. K., Kalinin, A. A., Do, B. T., Way, G. P., & Xie, W. (2018). Opportunities and obstacles for deep learning in biology and medicine. Journal of The Royal Society Interface, 15(141), 20170387.
17. Yan, X. Y., Zhang, S. W., & He, C. R. (2019). Prediction of drug-target interaction by integrating diverse heterogeneous information source with multiple kernel learning and clustering methods. Computational biology and chemistry, 78, 460-467.
18. Aleksandar Poleksic and Lei Xie (2018). Predicting serious rare adverse reactions of novel chemicals. Oxford Journals, Bioinformatics. 2018 Aug 15; 34(16): 2835–2842.
19. Kajal Negi, Arun Pavuri, Ladle Patel, Chirag Jain (2019). A novel method for drug-adverse event extraction using machine learning. Informatics in Medicine Unlocked. Elsevier, Science Direct, May 2019.
20. Alpha Vijayan, Dr. Chandrasekar B. S, "A Review on Various Approaches for Isolating Side Effects Causing Drug Elements", IJAST, vol. 29, no. 10s, pp. 4149-4158, Jun. 2020.
21. Dimitri, G. M., & Lió, P. (2017). Drugclust: A machine learning approach for drugs side effects prediction. Computational Biology and Chemistry, 68, 204–210.
22. Niu, Y., & Zhang, W. (2017). Quantitative prediction of drug side effects based on drug-related features. Interdisciplinary Sciences: Computational Life Sciences, 1–11.
23. Lo, Y. C., Rensi, S. E., Torng, W., & Altman, R. B. (2018). Machine learning in hemoinformatics and drug discovery. Drug discovery today, 23(8), 1538-1546.
24. Ding, P., Yin, R., Luo, J., & Kwoh, C. K. (2018). Ensemble Prediction of Synergistic Drug Combinations Incorporating Biological, Chemical, Pharmacological, and Network Knowledge. IEEE journal of biomedical and health informatics, 23(3), 1336-1345.

Non-Functional Requirements for Blockchain: Challenges and New Directions

Tahmina Akter Tisha and Mir Moynuddin Ahmed Shibly

Department of Computer Science and Engineering
East West University, Dhaka, Bangladesh

Email: shiblygnr@gmail.com

Abstract. Blockchain is one of the breakout technologies of the 21st century and it is destined to rule the field of cybersecurity. Eventually, all applications and software will adopt blockchain for being the most secure technology ever existed. Unlike other fields of computer science, blockchain too has functional and non-functional requirements (NFRs). As it is an emerging technology, Blockchain's non-functional requirements are yet to be well-defined. From requirements engineering perspectives, the NFRs for blockchain should be well-established. In this study, some of the major non-functional requirements of blockchain are thoroughly analyzed. And, challenges of NFRs for blockchain and new research directions have been proposed. Five non-functional requirements challenges are presented – lack of knowledge about NFRs for blockchain, the necessity of feasibility analysis, integration with software engineering, NFRs measurement metrics, and its formal language representation. Based on those challenges, four new research directions have also been proposed. If those challenges are dealt with properly, the world will surely march towards a safer future in terms of cybersecurity.

1. Introduction

Blockchain is one of the emerging technologies that is going to have a significant impact in the world where any sort of transaction happens. It is changing the way people interact with each other in terms of currency transactions. It is a system where a record of Bitcoin [1] or other cryptocurrency transactions is maintained through several computers that are connected to a peer-to-peer network. Blockchain technology is transforming the economy, finance, and money to the degree that its revolutionary power is being compared to that of the web and the internet in their early stages. As a consequence, all the development of software that revolves around blockchain technology is advancing at a spectacular rate [2]. As a revolutionary technology, it is expected to deliver substantial cost reductions by allowing transactions to be carried out directly between the users as peer-to-peer (P2P) processes. It provides an alternative to the conventional way of using a trustworthy third-party entity as an intermediary, for example, a financial institution like BBVA, Amazon, etc. Governments may use blockchain technology to deliver cybersecurity, automate processes, and incorporate hyperconnected services while improving trust and transparency. E.g., HACKMX, a project that uses blockchain technology to track and verify bids for public contracts, was launched in the digital government unit of the Mexican Ministry of Public Administration [3].

A blockchain is a kind of diary or spreadsheet that contains transaction information, and each transaction generates a hash. When the majority of the nodes approve a transaction then it is written into a block. Each block refers to the previous block and together make the blockchain. It is a growing chain of blocks where each block contains three fields – transaction info, participants involving in that transaction, and a cryptographic unique identifier i.e., a hash. Whenever a block is added to the chain, a copy of the block is stored in every computer that is connected to the network. Before adding a block to the chain, four criteria must have to be fulfilled – a transaction must have to take place, the transaction has to be verified, the transaction information has to be stored in a block and there has to a hash associated with the transaction. Blockchain is a decentralized ledger system with no central authority [4]. For being decentralized, it is impossible to alter a transaction whose information block is already added to the chain. Because, to do that, the majority of copies of the block of that transaction has to be altered to prove that the altered one is the real one. Unless other users will not accept that as the right one.

Although blockchain is impenetrable in terms of the security of a transaction, other requirements should be explored too. In conventional requirements engineering, the functionalities that a system must offer and must persistently provide are known as functional requirements [5]. Besides those functionalities, the system should be able to provide some other attributes like how the system will serve the functionalities. Such requirements are known as non-functional requirements (NFRs) [5]. For example, the software is meant to provide an interface between users and a bank so that the users can perform various transactions. Here, the process of secured transactions between parties is the functional requirements. On the other hand, the speed of transactions, the reliability of the service, the consistency, etc. are the non-functional requirements of that system.

There are many studies conducted for analyzing, exploring, and reasoning about the NFRs for regular software. These covered the most prominent NFRs like customer satisfaction, privacy, performance, usability, reliability, maintainability, and security [6][7]. As those software systems have various NFRs, a blockchain system should also have some NFRs and as blockchain is becoming one of the significant parts of secured transactions, it is worth exploring the NFRs for it and formulates a proper understanding of how a blockchain system behaves. This study aims to explore the challenges for NFRs of blockchain and derive some new directions from the perspectives of requirements engineering. However, the NFRs for the software systems are well-defined and well-understood. But when it comes to the blockchain-based system, the existing knowledge of NFRs may not apply. The existence of blockchain implies that the definition of certain NFRs for blockchain varies from typical software. The non-functional requirements are lesser known for blockchain-based systems. For example, what does it mean to be maintainable for a blockchain-based system? Do NFRs like flexibility and modularity remain relevant for the blockchain-based system? Some NFRs may become less important for a blockchain system while some may become more important. There are several common qualitative trade-offs in traditional NFR studies, also called conflicting NFRs

[7]. For example, there could be a trade-off between performance and privacy. The blockchain-based system needs to find out these types of conflicting trade-offs.

The aim of this study is to discuss whether the conventional understanding of NFRs and efficiency can be extended to blockchain-based systems from a requirements perspective. In order to do that, few research challenges and directions have been proposed related to the non-functional requirements of Blockchain technology. In this work, the justification has been presented that the conventional NFRs may need to be reorganized to cope up with the new technology like blockchain and the NFRs may need to be rethought or completely new techniques may need to be added to meet the paradigm change.

In the following section of this paper, state-of-the-art of blockchain technology with respect to non-functional requirements engineering has been described. After that, an example of blockchain in action has been illustrated. Then, the key research challenges of non-functional requirements for blockchain have been presented followed by the future directions. Finally, some concluding remarks have been presented.

2.Non-functional Requirements and Blockchain Technology

To make a system successful, it is very important to develop and consider the non-functional requirements properly [7]. On system can work just fine based on its functionality but can also fail for not handling the non-functional requirements (NFR) appropriately.

Non-functional requirements are product constraints or the features the system provides. They include constraints on timing, technology limits, and limitations imposed by standards. In certain cases, non-functional specifications refer to the device as whole systems or facilities, rather than individual device functions. This specific description, which describes something crucial in terms of what it is not, is not ideal, as many authors have discussed [8][9]. Our aim here is not to satisfactorily define NFRs but to explore their application to blockchain-based systems. NFRs can be treated as quality and the functional requirements can be recognized as an entity [10]. For example, a bus reservation system can be used to book seats using a secured transaction gateway. Here, booking the seats is functional requirements or entity of the system and security is the non-functional requirements or quality.

2.1 Qualities for Blockchain

Many works on blockchain have been described how blockchain works, how it is the most secure way to complete a transaction. These types of aspects are mainly analyzed and explained. Few works have been done regarding the non-functional requirements of a blockchain-based system. We are summarizing those works in this section.

UX Design. One unique obstacle to the creation of blockchain is the complexity of the user interface. The query is, how much of blockchain should you disclose to your consumers, as the enabling infrastructure? Do people need to know where they communicate with a blockchain anywhere on the application? Will they need to know what public and private keys are? How will it be kept secret if people do not know their private keys? Do they know about tokens, and need them to communicate with the application? These questions are relevant because people trust their money into your blockchain application. The value of the blockchain assets is usually backed by the value of a cryptocurrency coin, provided by an initial coin offering (ICO).

If the customer's wealth is connected to the coin's value, then how do you tell them their net worth increases and decreases as the value of the cryptocurrency increases and declines? When they are using your website, are they now investors in cryptocurrency? It is critical from the perspective of the user experience, too. Applications should be user friendly. For a new consumer, this first experience is very significant. Launching and cluttering a website with blockchain-related jargon is likely to put off most consumers. But dumbing it down can reduce the importance of it or the visibility of blockchain's main benefits that distinguish your application from the non-blockchain competition.

Scalability [11]**.** If the platform is connected to an implementation of a blockchain such as Ethereum then it is also connected to the scalability of the underlying network. Ethereum currently runs up to a theoretical limit of 25 (based on proof of work) at 4 transactions a second. When the next Ethereum game begins, a lot of transactions are created, would this affect the consumers? If they needed it, could they get their assets out immediately? Consumers would like to realize that they are not bound to something and will be able to return to the non-digital world at any stage of their choosing. To this, it is important to choose the right underlying platform.

Development Operations. As blockchain is still emerging, the technology to build a blockchain-based system is still evolving. It takes time to integrate new functionality into the system and improve the DevOps experience. Therefore, there should be enough amount of time associated with the development phase.

Managing Stakeholders. There is a tendency to over-sell the benefits of governance while marketing the advantages of a blockchain program. Possibly the actors of governance (e.g. police, government, auditors) get the most benefits from being able to monitor what everybody is doing and use it to control or test enforcement. It is necessary but other stakeholders need opportunities to enter the network otherwise there would be no users and no regulatory or control transactions. And there is the question of persuading all the stakeholders to get involved so you can do market research and collecting requirements. Modeling and smart contract concepts ensure that all various types of stakeholders recognize the company domains. It requires the views of several stakeholders to address the problem of what should be held in the blockchain [12].

Finally, who initiates and pays for the creation of the blockchain platform initiative? Blockchain's key advantage is that it can facilitate business and collaboration among organizations that have never met. Over the platform's lifespan, this would attract many organizations, but one company needs to take the initial step, see the benefit, and create the platform for everyone else to use.

3. NFRs for Blockchain Examples

In this section, we are going to present an example of how blockchain can be used in a real-life application. Suppose we are going to develop an application for an online retail shop a huge multinational company. This multinational company has a branch, where people can buy stuff even beyond the borders. For example, if a user from Bangladesh wants to buy an item from a seller from India, he/she has to pay the price of the product. Since the seller is abroad and there is no way to ensure the seller's authenticity physically and the seller cannot ensure the authenticity of the buyer either, a blockchain-based payment system can resolve this issue easily.

The borderless payment system is susceptible to mistakes, is costly, and is vulnerable to money laundering. Crossing the world takes days if not longer for the currency. The blockchain system can provide a solution to these problems. Many companies already adopted blockchain for remittance services like Abra and Veem s[13]. In 2004, Santander was one of the first banks to integrate blockchain into a payment app, enabling clients to make foreign payments 24 hours a day while clearing the next day and in 2019, they have launched the first-ever end-to-end blockchain bond [14]. These are just a few examples of using blockchain in the money transfer and banking sectors.

Now back to our example, if we want to ensure the trust persists between buyers and sellers, blockchain technology comes as the solution. Whenever a transaction happens, a new block will be added with the details of that transaction like date, time, the dollar amount in the first section of the block. In the second section of the block, there will be information about the participants of that transaction – in our case, the buyer from Bangladesh and the seller from India. Encrypted digital signatures of both parties will be there in the block and those digital signatures can be verified by a third party. Finally, the block will have a section consisting of a cryptographic hash. As the blockchain mechanism works, whenever the block will be added to the chain, millions of copies of that will be available online in an instance and the fraudulency in that transaction can never occur. In this way, the secure transaction between two parties that are thousands of miles away from each other and divided by borders can be secured and convenient. In this example, few non-functional requirements are needed to be considered. The developed application and interface for the money transactions should be user friendly. This NFR matches with the NFRs for a typical software. The other NFRs like managing stakeholders so that they do not have any confusion about the whole process should be acquired.

Another example of using blockchain can be in securing healthcare data. As we know, personal medical records and data are very sensitive and sophisticated. There is no alternative to securing and protecting those data. To ensure maximum security these data can be stored on the blockchain with a private key and the data will only be available to specific people who want to access the data after proper verification and validation. The same strategy could be used to ensure research is performed securely and confidentially through HIPAA laws [15]. Surgery receipts may be stored on a ledger and automatically submitted as evidence of delivery to the insurance companies. The ledger could be used for general health care administration, such as medication monitoring, compliance with legislation, test reports, and control of health care supplies too. For this secured medical records repository, the main NFR is security.

In both examples, we have discussed some common NFRs. But what if we add more blockchain-related NFRs to both systems like knowledge about the system. The users may not have an idea of how the system works, but they should be able to trust the system without any doubt. Other requirements like trust requirements [16], fairness, and transparency are the most important for a blockchain-based system. While developing these types of blockchain-based systems, these non-functional requirements should be implemented with the highest priority.

4. Research Challenges

Based on our illustrations, in this section of the article, few research challenges of non-functional requirements are presented.

Non-functional requirements identification challenge. The knowledge of NFRs acquired so far for blockchain is inconsistent and not complete, such as the method of describing and refining NFRs in the sense of blockchain specifics. For example, managing stakeholders is an important non-functional requirement for a blockchain-based system. How much knowledge a user knows about the system and how he/she perceives the system is necessary to understand to increase the usability of the system. The users have the idea that the system provides them the secured transaction, but are they trading off something for security? There are many questions to ask like this in the area of NFRs in the blockchain. These NFRs should be justified in blockchain contexts and other typical non-functional requirements should be refined so that those can be used in blockchain technology too. The NFRs specific to blockchain should be considered to use in other fields like software engineering, machine learning, the internet of things, etc.

Feasibility analysis challenge. Our knowledge about the feasibility of using blockchain in various systems is limited. Some studies explore the prospects of using this technology in different fields. Such Singh I and Lee S W [17] analyzed the feasibility of using blockchain for a secure cloud. To analyze the viability of using blockchain technology with the cloud, the authors in this study analyzed cloud and blockchain's security and trust Requirements and NFRs to satisfy cloud users' security and trust requirements. In another study [18], the feasibility and the sustainability of blockchain techniques in E-voting systems have been surveyed. This study demonstrated that blockchain technology can resolve issues with security concerns, integrity, anonymity, confidentiality, etc. The authors claimed and verified that the blockchain can ensure these attributes of E-voting systems. However, these are just two potential fields where blockchain can be used. There are other areas where blockchain fits and may perform well. Finding those and testing the feasibility of blockchain on them can be a challenging task.

Integration challenge. Another challenge in the field of studying NFRs for blockchain is to realize how it integrates with various software in terms of quality. Blockchain will have eventually be incorporated in every software which has some sort of payment system or has something that needs any sort of security for its betterment. A software engineer and a cybersecurity specialist or in our case a blockchain specialist have to work together to make an application that will serve its customers with maximum possible security. There are some examples where prospects of these two technologies working together have been analyzed.

Porru S et al. [2] have described the challenges and new directions of blockchain-oriented software engineering. The authors have explained some new challenges of implementing blockchain into software engineering and demonstrated a few new directions as well. In another study [19], an agile software development method for designing a blockchain application has been described. As software engineering is one of the primary disciplines and blockchain being one of the emerging disciplines in computer science, filling the gap between them can be pretty challenging.

Identifying measurement metrics challenge. As we have mentioned about the refinement of NFRs in the light of blockchain in our first challenge, we also need to understand the way to measure the non-functional requirements for blockchain-based systems. Working on software metrics in typical software engineering has established a range of potential metrics, connected to the required system NFRs, quantifiable at design or runtime e.g., to measure reliability by the number of runtime errors or mixed references measure modularity between components indirectly. These metrics are still applicable for blockchain-based systems, but are all the other measurement techniques still applicable? It is a question worth considering and deeper knowledge about it is desirable.

Non-functional requirements representation challenge. Researchers and stakeholders of blockchain lack a blockchain-specific way to articulate and define blockchain quality specifications, including goals and trade-offs, and the impact of domain background. While it is possible to reuse many of the current NFR-aware methods in a blockchain setting, we expect that the emergence of blockchain will expose several new concepts and challenges that alter the structural layout of extracting non-functional requirements for blockchain-based software. Formulating domain-specific ways to express those non-functional requirements can be helpful to understand the actual meaning of NFRs in the blockchain context.

5. New Research Directions

In the previous section, we have described five challenges concerning non-functional requirements for blockchain and we are acknowledging the fact that these are not the only challenges regarding the matter. There could be and must be many other challenges in the field of NFRs for blockchain. Now, in this section, we are going to present some research directions based on our proposed challenges. The directions that we are going to propose only reflect our thought process. The challenges can be solved by following other objectives too.

Obj1. To explore, identify, and define the non-functional requirements for blockchain. Under this objective, there is another sub-objective which is to refine already existing prominent NFRs for various fields like software engineering, so that some of those NFRs can be used in the context of blockchain. There are some ways by following which, this objective can be achieved. First, to know deeper insights about NFRs for blockchain, the specialists in this field should be asked through interviews, surveys, or any other data collection method. Secondly, a systematic literature review of all blockchain requirements can be conducted. And, finally, for the refinement purpose, the specialists from other fields can be asked and each domain-specific NFRs can be studied.

Obj2. In accordance with our second challenge, the second research direction is to conduct a feasibility study of blockchain technology in the areas where blockchain has not yet been implemented. If the feasibility study is conducted in every potential field where blockchain can be used, the new NFRs for it based on that specific scenario will come to light and the overall NFRs will be richer.

Obj3. The third objective is to analyze the gap between software engineering and implementing blockchain in software effortlessly. More studies should be conducted on how to use blockchain in an application on a general level. To make sure that blockchain will work along with software engineering, new frameworks can be developed.

Obj4. The final research direction in this study will cover the 4th and 5th challenge from the previous section. The challenges were to introduce the metrics for measuring NFRs for blockchain and to define the NFRs for blockchain in a domain-specific way. The direction to solve both challenges is to create, develop, and define a language representation to express the non-functional requirements for blockchain. The first step to develop this is to develop the semantics for the non-functional requirements concepts for blockchain appropriately. When the development of semantics is done, the graphical and textual syntax can be designed. In this way, a complete language representation of NFRs for blockchain can be developed. This well-documented representation can surely open many other prospects of NFRs in the blockchain-specific NFRs research.

6. Conclusion

This study was an attempt to find out more about the non-functional requirements for a newly emerged technology – blockchain. NFRs change and vary from one technology to another. Some change and some remain the same for all. In this paper, we tried to outline the challenges of NFRs study in a specific domain. And, based on those challenges, we proposed some objectives and new research directions. Those challenges and objectives have significant potential to look at and carry on further research from the requirement engineering perspectives. Conducting more research on this field of blockchain will surely lead us to a more secure and reliable user experience.

Acknowledgment

We are acknowledging the study titled "Non-Functional Requirements for Machine Learning: Challenges and New Directions" by Jennifer Horkoff [20]. Because we have conducted this study based on the pattern of that article.

References

[1] Nakamoto S 2008 A Peer-to-Peer Electronic Cash System *J. Gen. Philos. Sci.*

[2] Porru S, Pinna A, Marchesi M and Tonelli R 2017 Blockchain-oriented software engineering: Challenges and new directions *Proc. - 2017 IEEE/ACM 39th Int. Conf. Softw. Eng. Companion, ICSE-C 2017* 169–71

[3] Floyd D 2018 Mexico Tests Blockchain to Track Public Contract Bids *CoinDesk*

[4] Zheng Z, Xie S, Dai H, Chen X and Wang H 2017 An Overview of Blockchain Technology: Architecture, Consensus, and Future Trends *Proc. - 2017 IEEE 6th Int. Congr. Big Data,*

BigData Congr. 2017 557–64

[5] Sindhgatta R and Thonse S 2005 Functional and non-functional requirements specification for enterprise applications *Lecture Notes in Computer Science*

[6] Chung L, Nixon B A, Yu E and Mylopoulos J 2000 *Non-Functional Requirements in Software Engineering*

[7] ISO/IEC 2011 ISO/IEC 25010:2011 - Systems and software engineering -- Systems and software Quality Requirements and Evaluation (SQuaRE) -- System and software quality models

[8] Glinz M 2007 On non-functional requirements *Proceedings - 15th IEEE International Requirements Engineering Conference, RE 2007*

[9] Chung L and Do Prado Leite J C S 2009 On non-functional requirements in software engineering *Lecture Notes in Computer Science (including subseries Lecture Notes in Artificial Intelligence and Lecture Notes in Bioinformatics)*

[10] Li F L, Horkoff J, Mylopoulos J, Guizzardi R S S, Guizzardi G, Borgida A and Liu L 2014 Non-functional requirements as qualities, with a spice of ontology *2014 IEEE 22nd International Requirements Engineering Conference, RE 2014 - Proceedings*

[11] Croman K, Decker C, Eyal I, Efe Gencer A, Juels A, Kosba A, Miller A, Saxena P, Shi E, Gün Sirer E, Song D and Wattenhofer R 2016 On Scaling Decentralized Blockchains Initiative for CryptoCurrencies and Contracts (IC3) *Int. Conf. Financ. Cryptogr. Data Secur.* 106–25

[12] Hardwick P Non functional requirements and Blockchain *Capagemini*

[13] Levy A 2018 There's a lot of blockchain hype, but money-transfer start-up Veem is using it today *CNBC*

[14] Huillet M 2019 Santander Issues $20 Million End-To-End Blockchain Bond on Ethereum *Cointelegraph*

[15] DOH 2002 Department of Health and Human Standards for Privacy of Individually *Fed. Regist.* 1–93

[16] Khalifa D, Madjid N A and Svetinovic D 2019 Trust requirements in blockchain systems: A preliminary study *2019 6th Int. Conf. Softw. Defin. Syst. SDS 2019* 310–3

[17] Singh I and Lee S W 2018 Comparative requirements analysis for the feasibility of blockchain for secure cloud *Communications in Computer and Information Science*

[18] Çabuk U C, Adıgüzel E and Karaarslan E 2018 A Survey on Feasibility and Suitability of Blockchain Techniques for the E-Voting Systems *Ijarcce* **7** 124–34

[19] Marchesi M, Marchesi L and Tonelli R 2018 An Agile Software Engineering Method to Design Blockchain Applications 1–8

[20] Horkoff J 2019 Non-functional requirements for machine learning: Challenges and new directions *Proc. IEEE Int. Conf. Requir. Eng.* **2019-Septe** 386–91

Extracting information for failure prediction from intermittent data

Balakrishna S Kesavan[1] and Amol B Mahamuni[2]

[1] Principal Data Scientist, IBM India

[2] Program Director, IBM India

E-mail: balakrishna.kesavan@ibm.com

Abstract. The edge devices used for edge analytics usually record data intermittently. So, it is hard to extract patterns of device failures, using common temporal approaches, like Long Short-Term Memory, or multivariate models, like boosted trees. Survival analysis can produce actionable insights in such scenarios.

Key words. Edge analytics, survival analysis, failure prediction, predictive maintenance

1. Introduction

With Industry 4.0 gaining acceptance, there will be an increase in the number of intelligent devices, like Industrial IoT (IIoT) devices, that will be used at the edge locations of businesses. Many use cases for edge analytics have been identified at factories, retail stores, warehouses, hotels, distribution centers, and in vehicles. In all of these, the edge device itself will need predictive maintenance, so that the intended use case is reliably supported.

One challenge we face is that edge devices record observations intermittently. This could be because these devices are powered down during non-working hours or they may have limited storage space etc.

The challenge – intermittent data

To explain the problem and proposed solution better, we use S.M.A.R.T. (Self-Monitoring, Analysis and Reporting Technology) data about Hard Disk Drive (HDD) failures, hosted for public use by Black Blaze[1]. The SMART dataset has 3+Mn observations, covering 69 HDD models, of which only 215 observations show HDD device failures. The observations were recorded once a day starting from 1st Jan 2016 to 29th Apr 2016. There are 94 features (sensor readings), in addition to the 'failure' flag. The data contributors, Black Blaze, have found the following features, in Table 1, to be useful for failure prediction[2]. This is an important point of collaboration between device engineering and data science.

SMART 5		Reallocated Sectors Count
SMART 187		Reported Uncorrectable Errors
SMART 188		Command Timeout
SMART 197		Current Pending Sector Count
SMART 198		Uncorrectable Sector Count

Table 1. Features with engineering relevance to failure prediction.

To simplify this discussion, we subset the data by picking only 4 models, 'Hitachi HDS5C4040ALE630', 'WDC WD800BB', 'WDC WD800AAJS' and 'WDC WD3200BEKT'.

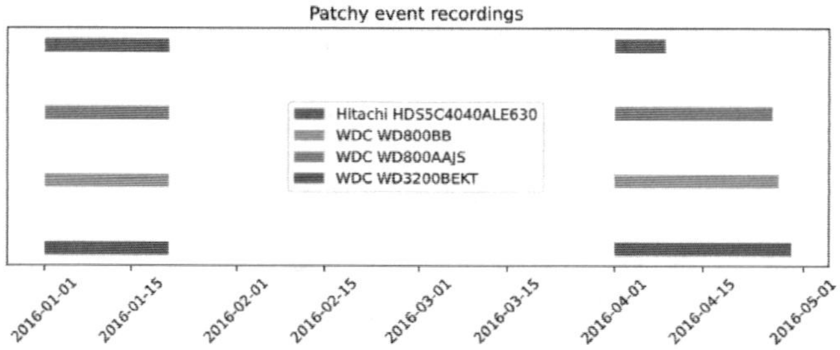

Fig 1.1. Intermittent observations

As seen, in Fig 1.1, data has been recorded in an irregular manner that varies by device. All devices' data have been recorded for a couple of weeks in January. Then, nothing is recorded till April when, data is recorded for some devices. But not for the same days.

2. Temporal analysis:

Temporal models, like Long Short Term Memory (LSTM), try to detect patterns of sensor readings that indicate device degradation and impending failure. For example, as the value of SMART_5_Reallocated Sectors Count goes up, the probability of device failure goes up[3].

But with the data being recorded intermittently for varying durations, as seen in Fig 1.1, it is impossible to even set a reasonable time 'window'/ timesteps (variable N in Fig 1.2) for an LSTM. For example, if we picked N = 14 days (refer Fig 1), it might work in the first couple of weeks of the dataset. But after that, data is missing for nearly 3 months. It would be wrong to impute missing data when the duration of missing data is longer than available data. This obstacle applies to other time series analysis techniques as well.

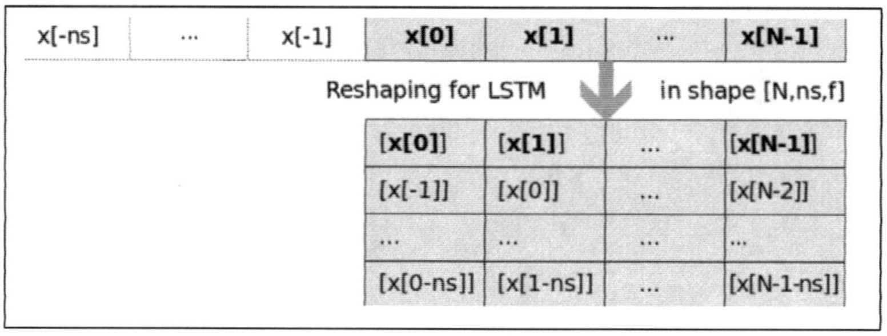

Fig 1.2. Reshaping timeseries data for LSTM

3. Multivariate analysis:

Multivariate models, like boosted trees and Cox Proportional Hazard were built. The full dataset was used, and the device model names were included as dummy features. The boosted tree showed poor fit with cross validated f1 score mean = 0.229, standard deviation = 0.043. (The target f1 score is 1.0). Imbalanced classes (very few failures) is the likely cause. Imbalanced class treatment techniques, like Synthetic Minority Oversampling Technique (SMOTE), could potentially inject fake patterns and were eschewed.

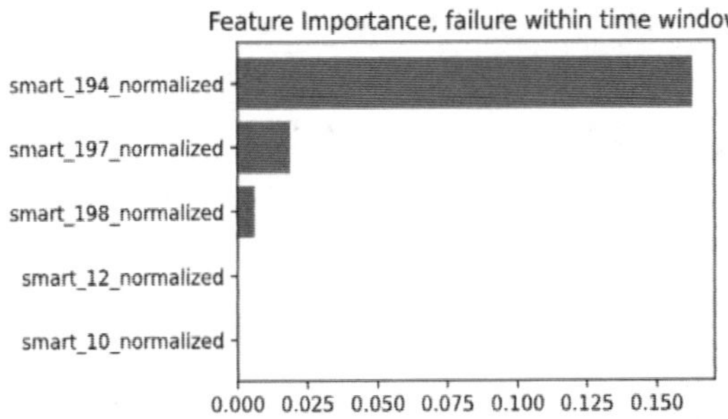

Fig 3.1. Near zero feature importance in boosted trees

The feature importance plot in Fig 3.1, from the boosted tree model, shows that even the top 5 features have very poor contribution, explaining the low f1 score. Only 2 of the features that Black Blaze suggested, for hard disk engineering reasons, appear here (smart 197 and smart 198). Smart 194 has 77 distinct values compared to 6 & 5 for smart 197 and smart 198 respectively. This is the likely reason for the inclusion of this feature, which was already flagged as non-informative by Black Blaze.

Fig 3.2. Correlation heatmap between suggested features to the time to failure (duration)

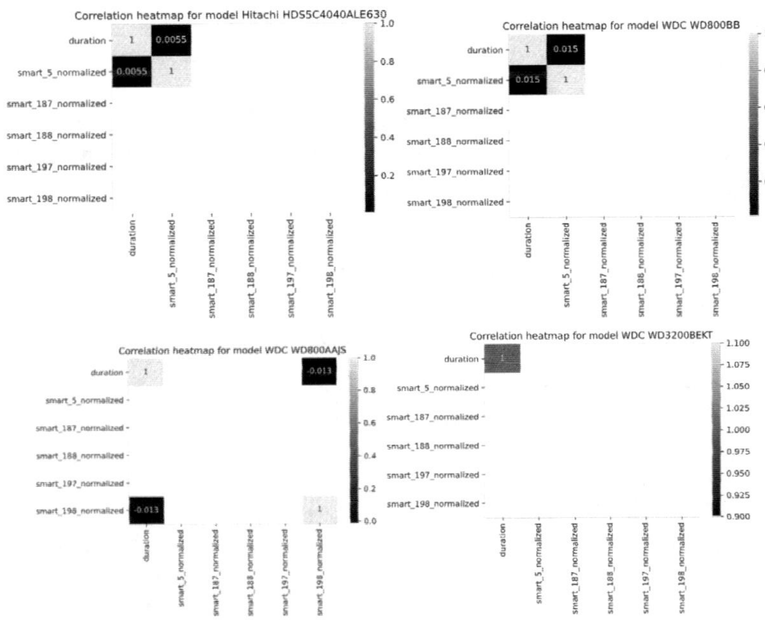

Survival regression using Cox Proportional Hazard was also tried but produced the error "ConvergenceError: Convergence halted due to matrix inversion problems. Suspicion is high collinearity". But as seen in Fig 4, there isn't any such collinearity. So, the more likely reason is that the features were not informative, as already seen in Fig 3.2.

Some information can be inferred from just the device failure observations in the data, as seen in Fig 1.5. For example, for the model "Hitachi HDS5C4040ALE630", there were 6 failures out 15,814 observations. Since 4 of these failures occurred on day zero (refer Table 7 for failure events), the median is 0 days and the mean value of age at failure was 3.5 days.

However, these values cannot be generalized to the devices that are yet to fail. We cannot expect more than half of all "Hitachi HDS5C4040ALE630" devices to fail on either day 0 or 4 (the median and mean values of failure duration).

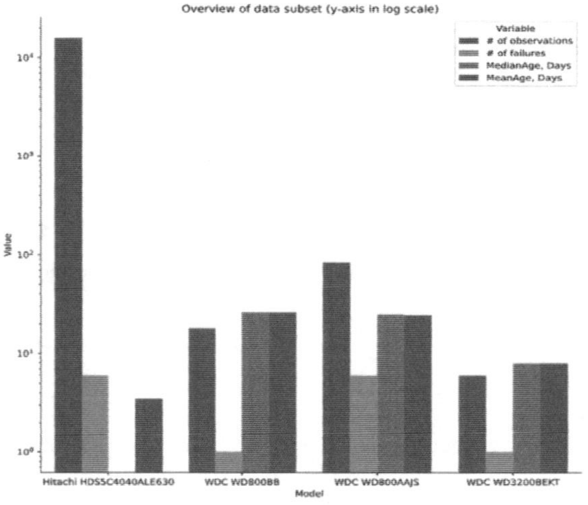

Fig 3.3. Overview of the data for four selected models (y-axis in log scale)

4. Squeezing out insights using survival analysis:

Using survival analysis, a univariate approach that uses the Kaplan Meier algorithm, we can still obtain usable insights out of this dataset.

serial_number	start	end	model	failure	duration
WD-WX71A9290300	2016-01-01	2016-01-01	WDC WD3200BEKT	0	0
WD-WX71A9290300	2016-01-01	2016-01-01	WDC WD3200BEKT	0	0
WD-WX71A9290300	2016-01-01	2016-01-21	WDC WD3200BEKT	0	20
WD-WX71A9290300	2016-04-01	2016-04-01	WDC WD3200BEKT	0	0
WD-WX71A9290300	2016-04-01	2016-04-01	WDC WD3200BEKT	0	0
WD-WX71A9290300	2016-04-01	2016-04-09	WDC WD3200BEKT	1	8

Table 2. Partially processed observations for the model "WDC WD3200BEKT"

Consider the partially processed observations for the model "WDC WD3200BEKT" in Table 2. There are only 6 of them in the dataset, of which only the last observation shows a failure.

```
WDC WD3200BEKT:
          removed  observed  censored  entrance  at_risk
event_at
0               4         0         4         6        6
8               1         1         0         0        2
20              1         0         1         0        1
```

Table 3. Survival for the model "WDC WD3200BEKT"

Using the observations from Table 2, a survival table, in Table 3, extracts information about how many devices of this model were being tracked ("at risk") at different points in time ("event_at", in days), how many of these recorded no failures during or at the end of the study ("censored") and how many failed ("observed"). Survival analysis uses information even from devices that never failed and hence, were "censored". Instead of discarding observations relating to devices that didn't fail, censoring allows using all available information. That is, the fact that all 6 devices survived day 0 and 1 device has survived ***past*** day 20 is used in survival probability computation.

The survival probability[4, 5] is calculated using this formula where 'n' is the number of devices at risk ('at_risk') just before time 't'('event_at') and 'd' is the number of deaths ('observed') at 't'.

$$\hat{S} = \prod_{t_i<t} \frac{n_i - d_i}{n_i}$$

Day	Number of devices at risk	Number of devices that failed	Survival probability (product of terms)
0	6	0	(6-0)/6 = 1
8	2	1	{(6-0)/6}x{(2-1)/2} = 0.5
20	1	0	{(6-0)/6}x{(2-1)/2}x{(1-0)/1} = 0.5

Table 4: Survival probability calculations (Refer Table 3 for the variable values)

It is important to note that censoring is assumed to be non-informative and that censored devices are not excluded because of increased risk of failure.

It is also possible to extrapolate past our observed durations by using parametric models, like Weibull. But this is not covered here.

Insight 1 – survival probabilities for a device model: The survival function plot, in Fig 6, provides an easy to interpret view of survival probabilities. For example, for model "WDC WD3200BEKT" the probability of survival is 50% on the 10th day. Using this information, predictive maintenance schedules can be set. For example, inspect/ maintain/ replace components before probability of survival drops below 50%.

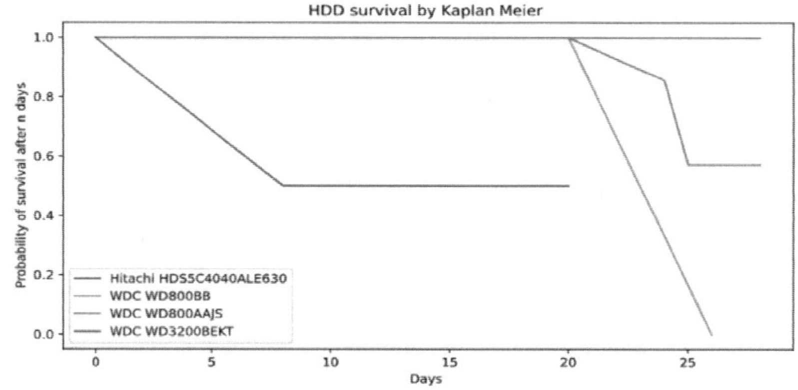

Fig 4.1. Survival function plot showing survival probability over time.

Insight 2 – comparing device models at a future point in time: The Restricted Mean Survival Times (RMST), shown in Fig 4.1, another measure from survival analysis, is the area under the survival curve up to a prespecified time horizon.

All else being equal, the model with the larger RMST has the higher probability of surviving at a given point in time (day 28 in this case). While the RMST for "WDC WD800BB" and "WDC WD800AAJS" are deceptively close, the plot shows that some "WDC WD800AAJS" devices are still running whereas no more "WDC WD800BB" devices are. Cross check with details in Table 7. If these device models were equivalents, the expected duration of operation before testing/ replacement can help choose the model with lower maintenance costs. (Use a log rank test to test if two survival curves are statistically different).

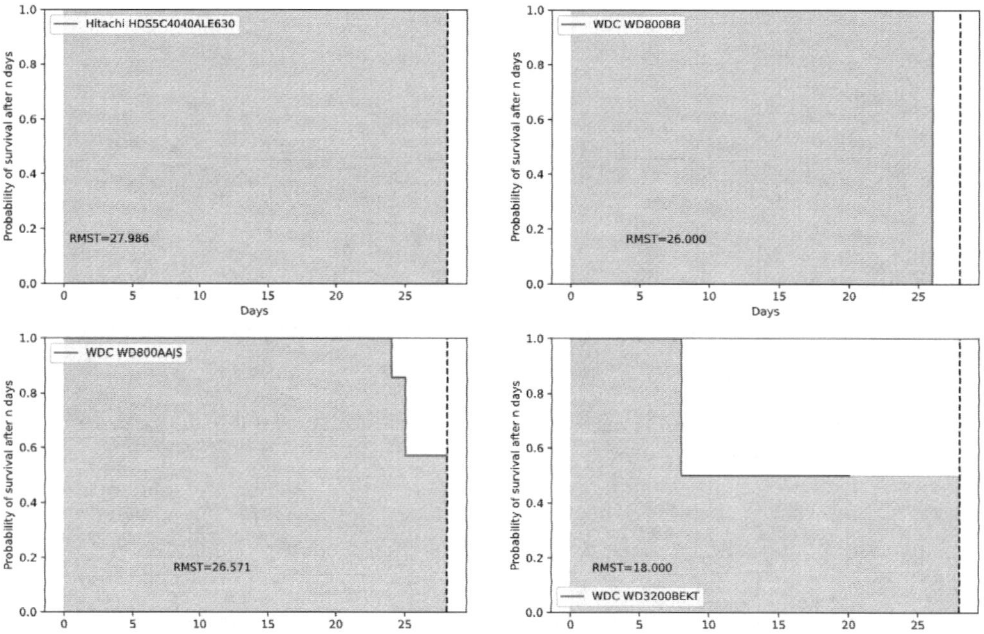

Fig 4.2. Comparing survival of models at a future point in time

Insight 3 – Confidence in estimates. Continuing the above comparison between "WDC WD800BB" and "WDC WD800AAJS", the 95% confidence interval around the median survival time is instructive. We can tell for sure that more than half the "WDC WD800AAJS" devices were running on day 26. But we can't say the same for "WDC WD800BB". See Table 5.

0	Lower CI	Median Survivial Time, Days	Upper CI
Hitachi HDS5C4040ALE630	inf	inf	inf
WDC WD800BB	26.0	26.0	26.0
WDC WD800AAJS	25.0	inf	inf
WDC WD3200BEKT	8.0	8.0	inf

Table 5. Median survival time from survival analysis

But by considering only the failure observations, in Table 6, we underestimate the median & mean ages. As stated earlier, we cannot expect more than half of all "Hitachi HDS5C4040ALE630" devices to fail on either day 0 or 4 just because 6 out of 15,814 devices failed by then.

model	# of observations	# of failures	MedianAge, Days	MeanAge, Days
Hitachi HDS5C4040ALE630	15814	6	0.0	3.500000
WDC WD800BB	18	1	26.0	26.000000
WDC WD800AAJS	84	6	25.0	24.666667
WDC WD3200BEKT	6	1	8.0	8.000000

Table 6: Median survival time from failure observations

In the case of model "WDC WD800BB" of 18 devices, 17 were censored and at the end, 1 failed on the 26th day. Hence the single median survival time estimate. In the case of "WDC WD800AAJS", since less than half the number of devices at risk at any point in time failed, the median cannot estimated. Refer Table 5 and Table 7. For "WDC WD320BEKT", since half the devices at risk have failed, a median can be estimated but since one more is still running, upper bound cannot be estimated.

5. Conclusion

We saw that due to the intermittent data, temporal analysis approaches could not be used to extract patterns of failure. We also saw that multivariate approaches, like boosted trees survival regression are not helpful when the classes are imbalanced, and the features are not informative. However, survival analysis is able to provide actionable insights, when other methods cannot.

These are the events,by model, by time (column "event_at"):

****

Hitachi HDS5C4040ALE630:

event_at	removed	observed	censored	entrance	at_risk
0	10544	4	10540	15814	15814
1	1	1	0	0	5270
20	2638	1	2637	0	5269
26	1	0	1	0	2631
28	2630	0	2630	0	2630

WDC WD800BB:

event_at	removed	observed	censored	entrance	at_risk
0	12	0	12	18	18
20	5	0	5	0	6
26	1	1	0	0	1

WDC WD800AAJS:

event_at	removed	observed	censored	entrance	at_risk
0	56	0	56	84	84
20	14	0	14	0	28
24	2	2	0	0	14
25	4	4	0	0	12
28	8	0	8	0	8

WDC WD3200BEKT:

event_at	removed	observed	censored	entrance	at_risk
0	4	0	4	6	6
8	1	1	0	0	2
20	1	0	1	0	1

Table 7. Full survival table for the four selected models

6. References:

[1] Data source: https://www.kaggle.com/backblaze/hard-drive-test-data

[2] Black Blaze feature recommendation: https://www.backblaze.com/blog/what-smart-stats-indicate-hard-drive-failures/

[3] Source: https://en.wikipedia.org/wiki/S.M.A.R.T.

[4] "Survival Analysis A Self-Learning Text" by David G. Klein Baum and Mitchel Klein, published by Springer.

[5] Python Lifelines: https://lifelines.readthedocs.io/en/latest/index.html

Data analysis of a match using soccer-log

R Srinath[1], **NagaSwetha Devarakonda**[2], **Arun Biradar**[3] **and Shilpa Biradar**[4]

[1] Department of IS&E, NIE, Mysuru , Research scholar, EWIT (VTU), Karnataka,
[2] Managing Partner, Entelika Consulting and IT Services, Mysuru, India,
[3] Department of CS&E, EWIT, Bengaluru, Karnataka.
[4] Department of IS&E, Dr AIT, Bengaluru, Karnataka

E-mail: rampursrinath@nie.ac.in, SAnshu@Entelika.com, hodcsea@gmail.com, biradarshilpa@gmail.com

Abstract. Applying Machine learning, data analytics etc have proven a greater improvement in decision making in all levels. A special attention has been made in the field of sports. As a whole, team improvement is a main concerned in all directions. The goal of any team would be to win every match they play. This can happen only if there is an excellent team efforts i.e. good coordination between players. We propose Sports Analysis an approach for better performance day by day. In this paper we talk about acquiring the details of a match and presenting team statistics.

Keywords: Data Analysis, Soccer, Machine Learning, Soccer-log.

1. Introduction

Machine learning, Artificial Intelligence has been making a foot print in the current market either education, manufacturing, marketing or any field to mention. Even it has made its way in sports also. Sports Analytics is a new mention that is happening in the current world. Every player, coach or club objective is to win every match they play. All of them have the will to win but important is the preparation to win. Machine learning is assisting all kinds of sports to get out of all flaws and to strengthen their capabilities.

Now there is a tremendous growth in applying different Machine Learning algorithms to all kinds of sports. This has increased the performance as a whole to the team and club. Soccer has been a wonderful sport and that most watched one by the crowd. Soccer analytics has gained lot of importance in today's world. In this paper, we specifically talk about soccer-log and evaluations of performance of a match.

2. Related Work

[1] Aimed at study of current state in analytics (soccer) and predicted the outcome of soccer match which would benefit for sport analysts, managers of soccer clubs, coaches. Soccer statistics are attracting wide interest from a long time ago. [2] Focused on the most challenging open problems in soccer analytics.

Data-driven analysis has been made a tremendous change in team sport. [3] Useful of visualizations has been made to good extent so that many challenges have been resolved. [4] Position tracking had been used in this paper to unlock what Sports analytics can suggest.

Soccer has become a big dollar industry, which is a globally adapted sports. [5] Predicted the future of upcoming soccer matches. [6] The most influential performance metrics of the players are considered for determining the match outcome

Performance indicators are the key to be considered for sports that are related to attacking play in football. [7] Uses the above mentioned keys for providing appraisal of a player using machine learning concept.

Promising results have been seen using Machine learning (ML) particularly in the area of prediction and classification. [8] Provides a sport results using Artificial Neural Networks (ANN) in terms of win or lose also provides focus on analysis of Machine Learning.

Sports Analysis is rapidly growing area of sports science with the ever increasing easy internet accessibility and recognition of Machine Learning. [10] Predict win or lose that is the outcome of the matches in English Premier League

[11] Proposed a procedure for trying to outcome as win or lose of a soccer match. Bayesian Model based on rank position is used along with a Machine Learning Model. It is modeled to work based on historical data of matches.

3. Sports Analysis

Sports Analysis is one stop solution for all the requirements of sports.

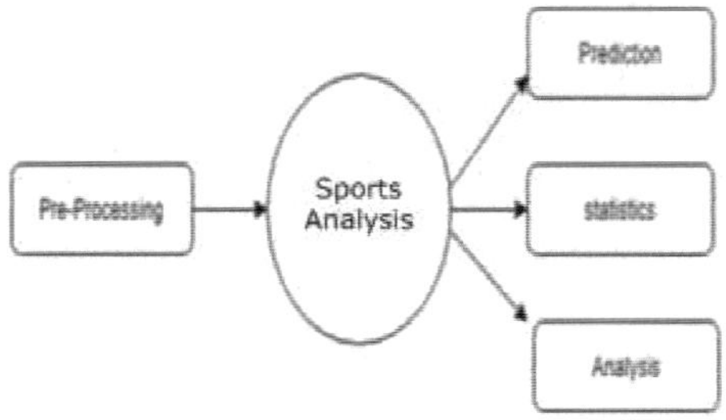

Figure 3.1: Data Analysis in Sports.

3.1 Pre-Processing

Pre-processing in this case is tagging which is achieved by using proprietary software. The input to tagging process is Video recording, which contains video tracking data that describes the player trajectories. The output of tagging is soccer-logs which gives the details of the events that occurred during a match.

Figure 3.2. Match tagging.

The Tagging of a match is a three step process.

Step 1: Setting formations.

Before starting tagging, the teams are set to analyze, Identify the own team and the opponent team and also the Jersey number of the corresponding players.

Step 2: Event tagging.

A new event on the timeline is created for each ball touch; the operator selects one player identifies using jersey number. Identify the type of the event and attach the type (e.g., pass, dribbling, tackle, shot, etc.) and subtype (e.g., Successful, unsuccessful, goal, saved, off target etc) to that particular player. There by the event is inserted in the streamlined way so that events are attached to the timeline. Normally special customized keys are used to attach the event ie short-cut keys

IOP Conf. Series: Materials Science and Engineering 1110 (2021) 012018 doi:10.1088/1757-899X/1110/1/012018

Step 3: Quality control.
One of the most important steps in tagging is the quality control, performed for each match. Quality control is done by selecting the customize tag and run the video with these customize tags manually and check for correctness. The threshold that has been set approval is 90% in each of these events.

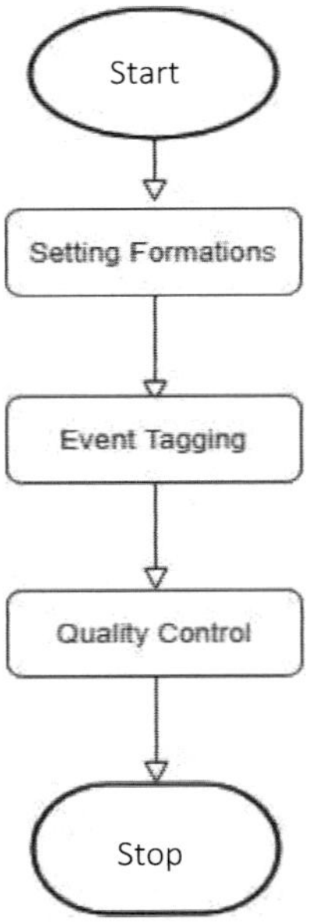

Figure 3.3. Steps in tagging.

The events captured in the soccer-log are Pass (successful, unsuccessful), Dribbling (successful, unsuccessful), Tackle (successful, unsuccessful), Safe goal, Shot (Goal, saved, off target), Back post pass, Possession etc

3.2 Soccer-log
Soccer-log is the output of tagging process, which contains the event details that occurred during a match. Soccer-log is an XML file with event details. The event Possessions sample in soccer-log is of the form

```
<instance>
<ID>699</ID>
<start>640.465</start>
<end>644.05</end>
<code>Team1  Possessions</code>
<label>
<group>Possessions</group>
<text>Team Possession</text>
</label>
<label>
<group>Set Pieces</group>
<text>Goal Kick</text>
</label>
</instance>
```

Figure 3.4. Event representation in Soccer-log.

4. Analysis using soccer-log

The output of tagging is the Soccer-log; using the events in the log file the match statistics are derived. The match statistics are as below:

4.1 Shot Statistics:

The shot is one of the major attribute for a team to win and the shot statistics shows the team effort. In spite Team2 has won, the statistics show a different scenario, out of 33 Shots, only 10 shots were goal so efficiency is 30% which is not actually a good one for a threshold 60%. The shaved shots of Team2 are 45%, so approximately 50% of the time Team1 was successful in blocking Team2 from making a goal.

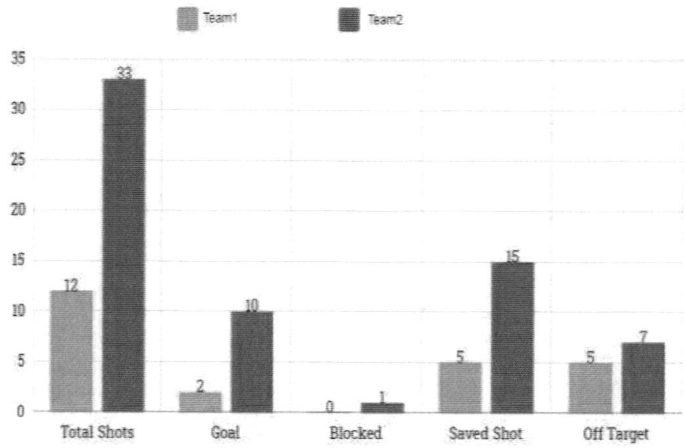

Figure 4.1. Shot Statistics

4.2 Set Pieces:

The set pieces snapshot give us a more understanding on how the team's performance has made a difference in the game. Free Kicks indicator shows that Team2 has done 12 fouls, which could have been a major disaster in case Team1 had taken advantage of this.

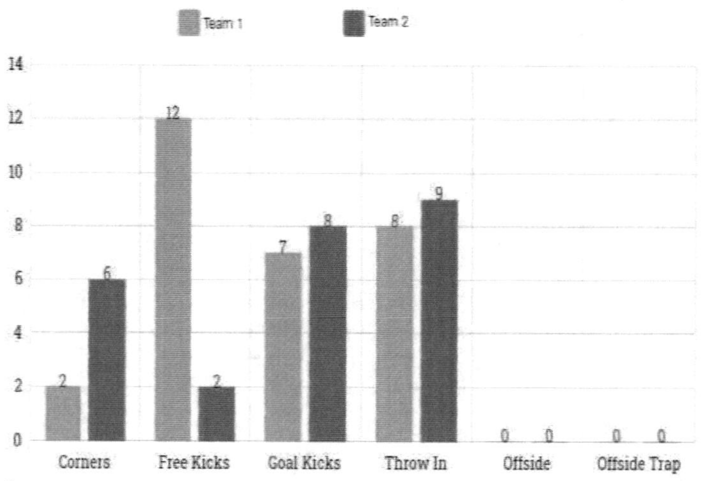

Figure 4.2. Set Pieces

4.3 Possession and Rating:

The output of tagging is the Soccer-log; using the events in the log file the match statistics are derived. The match statistics are as below:

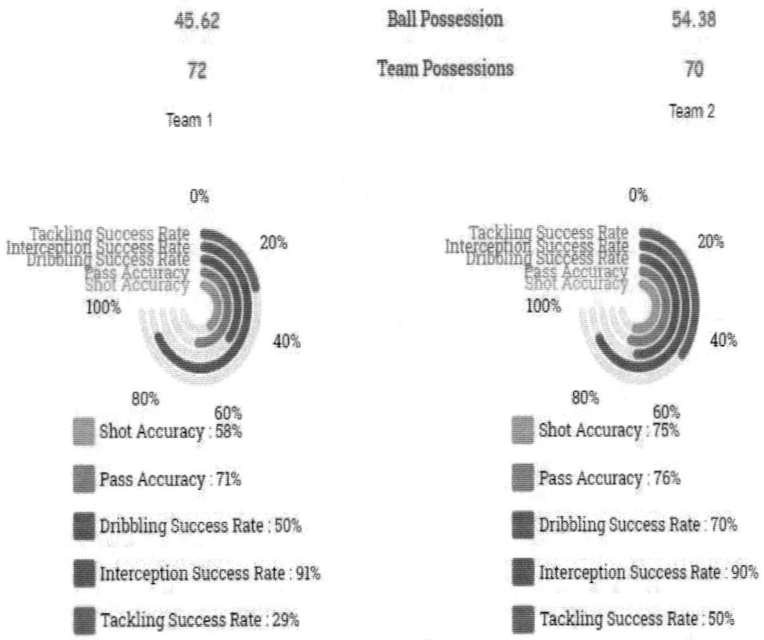

Figure 4.3. Possession and Rating.

4.4 Team Play:

The output of tagging is the Soccer-log; using the events in the log file the match statistics are derived. The match statistics are as below:

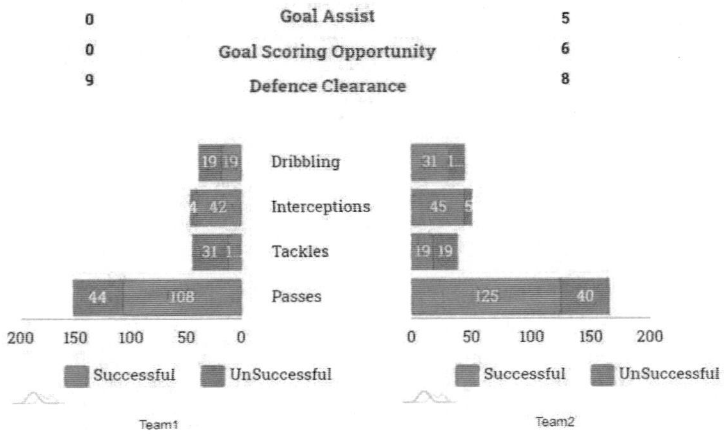

Figure 4.4. Team Play.

IOP Conf. Series: Materials Science and Engineering 1110 (2021) 012018 doi:10.1088/1757-899X/1110/1/012018

5. Conclusion

Sports Analysis help understands the insight of a match in more detail which helps in developing more skills for a player. It engages with game footage, a safe communication platform and provides the data in the form of statistics of a match, which help for further actions to be taken.

6. Reference

[1]. Sukumar, Sunil Srinivas, "Soccer Analytics & its future" (2019). Creative Components. 254. https://lib.dr.iastate.edu/creativecomponents/254.

[2]. Pappalardo, Luca. (2019). Soccer Analytics: how Data Science is changing the "Beautiful Game". 10.13140/RG.2.2.14459.98088.

[3]. Stein, Manuel & Häußler, Johannes & Jäckle, Dominik & Janetzko, Halldor & Schreck, Tobias & Keim, Daniel. (2015). Visual Soccer Analytics: Understanding the Characteristics of Collective Team Movement Based on Feature-Driven Analysis and Abstraction. ISPRS *International Journal of Geo-Information*. 4. 2159-2184. 10.3390/ijgi4042159.

[4]. F R Goes, L A Meerhoff, M J O Bueno, D M Rodrigues, F A Moura, M S Brink, M T Elferink-Gemser, A J Knobbe, S A Cunha, R S Torres & KAPM Lemmink (2020) Unlocking the potential of big data to support tactical performance analysis in professional soccer: A systematic review, European Journal of Sport Science, DOI: 10.1080/17461391.2020.1747552.

[5]. Berrar D, Lopes P,Davis J Guest editorial: special issue on machine learning for soccer. Mach Learn 108, 1–7 (2019). https://doi.org/10.1007/s10994-018-5763-8

[6]. Kumar, Gunjan. (2013). Machine Learning for Soccer Analytics. 10.13140/RG.2.1.4628.3761.

[7]. Herold, Mat & Goes, Floris & Nopp, Stephan & Bauer, Pascal & Thompson, Chris & Meyer, Tim. (2019). Machine learning in men's professional football: Current applications and future directions for improving attacking play. *International Journal of Sports Science & Coaching*. 14. 10.1177/1747954119879350.

[8]. Rory P, Bunker, Fadi Thabtah," A machine learning framework for sport result prediction", *Applied Computing and Informatics* Volume 15, Issue 1, January 2019, Pages 27-33.

[9]. Hossam M Zawbaa1 , Nashwa El-Bendary2 , Aboul Ella Hassanien1 , and Tai-hoon Kim3, "Event Detection Based Approach for Soccer Video Summarization Using Machine learning", *International Journal of Multimedia and Ubiquitous Engineering* Vol. 7, No. 2, April, 2012.

[10]. Abha Tewari, Tushar Parwani, Ajinkya Phanse, Akshay Sharma and Anush Shetty. Soccer Analytics using Machine Learning. *International Journal of Computer , Applications* 181(50): 54-56, April 2019.

[11]. L Hervert-Escobar, T I Matis and N Hernandez-Gress, "Prediction Learning Model for Soccer Matches Outcomes," 2018 Seventeenth Mexican *International Conference on Artificial Intelligence (MICAI)*, Guadalajara, Mexico, 2018, pp. 63-69, doi: 10.1109/MICA I46078.2018.00018.

A Comparative Analysis of Pre-Processing Time in Summary of Hindi Language using Stanza and Spacy

Atul Kumar[a],Dr Vinodani Katiyar[b],Dr Pankaj Kumar[c]

[a]Research Scholar,DSMNRU Lucknow,India, atulverma16@gmail.com

[b]DSMNRU Lucknow,India, drvinodani@gmail.com

[c]SRMGPC,Lucknow,India, pk79jan@gmail.com

Abstract. Text summarization is a method or say way for converting large texts into smaller ones by keeping all important points of the larger text as it is in the smaller ones and giving the output in the modified form called summary of original text document. This task is very difficult for humans as it required large amount of rigorous analysis of the document. In this proposed paper we are comparing pre-processing time for two tools of natural language processing one is STANZA and the other is SPACY both are based on modern technologies and are examine for HINDI language processing. In this paper we are preforming a comparative study of both the tools on the bases of their pre-processing time for processing HINDI language.
Now a days, a lot of text summarizer tools are present in the environment and we are keen to use them but here the point of this paper comes into light that how we know which Summarizer is fast enough to get us the same accuracy.

Keywords— Stanza, Spacy, Pre-processing time, Summarizer, Natural Language Processing, Tokenization, Segmentation, Stemming, Lemmatization.

1.Introduction

In this present scenario information is very important thing exists. Billions of data is floating on the internet every second, but that data also includes a lot of information which are not important in order to make it further use in any domain so, as a solution to this problem TEXT SUMMERIZER came into picture, it helps the people to get the information they required at that time by eliminating unwanted words in that text or document. Text summarization is used by a lot of applications like for instance, scientists require a tool to provide summaries for figuring out whether or not to study the entire record(text) or now no longer and for summarizing facts searched through person on Internet. News corporations can use different document summarizer to group the data from various sources and summarize that. In this proposed paper we are providing a tool by which the user can get the fast output so that it can increase the productivity of their domain.

2. The Related Work

There exist a great deal of summarizers over the web, yet a significant number of them uphold just two dialects that are English and European dialects. The first of the main work in robotized text rundown was given by Luhn [1] in 1958. To produces outlines he utilizes frequencies of words by given regarded contribution with the item to decide the contiguity of a sentence in an info.

P.B. Baxendale [2] in 1958 proposed a novel element that is sentence area or sentence area position in an info report. It is investigated that sentences which are situated toward the start or toward the finish of the report are more significant than different sentences in the report. A sentence position include turned into a basic component for sentence extraction and it is utilized till now.

H.P. Edmundson [3] in 1969 proposed a novel structure for text rundown. He proposed two new highlights. To begin with, prompt words that are the presence of most demonstrative words into a archive, for example, at long last, in a rundown, ultimately, and so forth Second, title or heading Words that is an extra weight was allocated to a sentence, if sentences have heading words in it.

Afterward, Julian Kupiec [4] in 1995 proposed some new highlights like sentence length, presence of capitalized words,express structure, and furthermore included different highlights created beforehand. Likewise, portrayed another strategy of the rundown with an innocent Bayes classifier, the grouping capacity orders each sentence regarding if it is extraction commendable.

Gupta Vishal and Gurpreet Singh Lehal [5] in 2010, In which extractive summarization method consists of selecting important sentences, paragraphs etc. from the original document and concatenating them into shorter form .

Jawline Yew Lin [6] in 1999 additionally attempted to show a framework for sentence extraction utilizing choice trees for making useful conventional/question situated concentrates. Jawline Yew Lin recognized various highlights, for example, : title, pattern, position, tf-idf, question signature, sentence length, lexical network, citation, first sentences, formal person, place or thing, pronoun and descriptor, mathematical information, non-weekend days, and month. The score for all highlights was joined with programmed learning utilizing choice trees and blend work. Lin directed a profound investigation of various highlights impacts through a glass-box. The test result shows that there was not the single component which performs well for question-based rundowns. Highlights like mathematical information give better outcomes for the question which requires an answer in mathematical worth, while non-weekend days and month include gives the best outcome for inquiries like "when?"

Conroy and O" leary [7] made a framework for sentence extraction utilizing two procedures named QR and HMM. In QR procedure the significance of the sentence was determined and the sentences having higher significance were added to the rundown. At that point, an overall score of the excess sentences was modified on the grounds that a portion of the leftover sentences was repetitive. Furthermore, the other procedure was the concealed Markov model (HMM) which is a consecutive model for programmed text outline. What's more, in HMM just three highlights were utilized: sentence position, complete no of terms in the sentence, and the likeness of sentence terms in the given archive. Eventually, the assessment was finished by looking at HMM produced outline with the human-created rundown.

S. P. Yong et al [8] in 2005 portrayed a computerized framework that had the capacity of learning by joining diverse later approaches, for example: a factual methodology, sentences extraction and neural organization.

Ladda Suanmali et al [9] in 2009 proposed a framework based on the fluffy rationale. Fluffy guidelines and fluffy sets were utilized for separating huge sentences dependent on sentence highlights.

Different highlights were utilized to register sentence importance and the worth was given somewhere in the range of „0" and „1" if a specific element was available in the sentence. The element utilized was title word in a sentence, sentence centrality, likeness to initially sentence, watchword, sentence length, sentence position, formal person, place, or thing and numeric information. The estimations of these highlights were given as a contribution to a fluffy framework, which made a yield dependent on highlights and IF-THEN principles characterized by sentence extraction.

Li Chengcheng [10] in 2010 proposed a novel ATS strategy in light of Rhetorical Structure Theory (RST). RST is the documentations of expository connection introduced between two non-overlapping writings called the core (N) and satellite (S). The trial perceptions were utilized to communicate the distinction among N and S, which was communicated by N that what is more imperative to the writer's reason than S and the Logical core was intelligible autonomous of the S. The whole content was separated into little units called sentences dependent on delimiter like a full stop, commas, any accentuation mark found between sentences.
The whole cycle was based on a diagram, sentences which were less significant was taken out from the diagram and remaining sentences were summed up.

S. Harabagiu and F. Lacatusu [11] propose a method that, given a catchphrase question, on the fly creates new pages, called formed pages, which contain all inquiry watchwords. The created pages are produced by getting and assembling related pieces from hyperlinked Web pages and holding connects to the first Web pages. To number the created pages, we consider both the hyperlink structure of the first pages and the relationship between the catchphrases inside each page. Besides, they introduce and tentatively assess heuristic calculations to productively create the top formed pages.

In R. Varadarajan, V. Hristidis, and T. Li [12] proposed an assessment mining and outline strategy utilizing various methodologies and assets, assessing every one of them thusly. Their work incorporates the improvement of the extremity arrangement segment by utilizing AI over-explained corpora and different strategies, for example, an anaphora goal.

In E. Lloret, A. Balahur, M. Palomar, and A. Montoyo [13] present the main report of programmed estimation rundown in the legitimate space. This work depends on preparing legitimate inquiries with a framework containing a self-loader Web blog search module and Fast Sum, a total programmed extractive multi-record assumption synopsis framework. They give quantitative assessment aftereffects of the rundowns utilizing legitimate master analysts. They report benchmark assessment results for question-based opinion synopsis.

In J.G. Conrad, J.L. Leidner, F. Schilder, and R. Kondadadi [14], the creator proposed a novel calculation for sentiment outline that assesses substance and cognizance, all the while. They think about a synopsis as a succession of sentences and straightforwardly get the ideal grouping from numerous survey archives by pulling and requesting the sentences. They accomplish this with a novel Integer Linear Programming (ILP) definition. The framework in this paper is a ground-breaking combination of the Maximum Coverage Problem and the Traveling Salesman Problem and is generally pertinent to message age and outline.

In H. Nishikawa, T. Hasegawa, Y. Matsuo, and G. Kikui [15] writers characterize an assignment called subject life structures, which sums up and relates the centrepieces of a theme transiently with the goal that per users can comprehend the substance without any problem. The proposed model, called TSCAN, determines the significant subjects of a point from the eigenvectors of a fleeting square affiliation framework.

3. Types of Text Summarization

We have two types of text summarization which are as follows:

3.1 Extractive Summarization:

Extractive summarization can be proposed as a classification subject. Its main target is to take out the most relevant sentences and paragraph from that respective text or data and ranked them high. The highest ranked regions from all text (data) can then combined and re-rank using same aspects and append them into smaller form. Extractive summarization uses statistical way or say technique to select important sentences or keyword from text (data).

Extractive summarization involves mentioned points: -

A. Pre-processing step
B. Processing step.

Pre-processing is a well arranged representation of the original document. It involves three sub processes:

1. **Sentence segmentation**: In this method sentence's boundary are find out and it is find out with the presence punctuation marks.

2. **Stop-Word Removal:** Stop-words and the words which do not provide relevant information to the subject are removes.

3. **Stemming:** Reason for stemming is to get the root word by eliminating prefix and suffix.

In processing phase, characteristic influencing the significance of sentences are decided and calculated, and after that weights are assigned to these characteristics using weight learning method. Final score of every sentence is calculated using Feature-weight equation and those sentences are chosen for the final summary whose scores are highest among all the sentences in the original document.

3.2 Abstractive Text Summarization:

Abstractive text summarization provides the summary after rigorous analysis of the provided document and design the summary using only required words and eliminate the rest of the unwanted words in order to make the text sort and easy to read and fast to understand.

In this method summarization of every text of the sentence is implemented and have different way as compared to the original document.

Abstractive summarization method can be divided into two approaches:

A. Structured approach
B. Semantic approach.

Structure approach uses various techniques to give the required result that are tree technique, template technique, ontology technique, lead and body phrase technique and rule technique.

Semantic technique uses techniques that are Multi-modal Semantic model, Information item technique, and semantic graph technique.

[1] Summarization Technique

As we are performing Abstractive method in pre-processing time-based comparative study on STANZA and SPACY here are the common techniques which both tools follows:

a. Pre-Processing:

In the pre-processing phase of these tools, the text is firstly divided into list of sentences and then, that sentences are additionally divided into words and after that common words are eliminated.

Pre-processing method includes three steps:

1) Segmentation
2) Tokenization
3) Stop words elimination.

1. Segmentation:

Here sentences are segmented, based on sentence boundary which can be predefined or as well as user defined too. In Hindi language, sentence boundary is identified by "|" which is known as full stop in English language. On every full stop, the sentences are fragmented and place into a list. The final yield of this phase is the list of sentence and this list is send for next level processing.

2. Tokenization:

In this phase, all the sentences are divided into words. In Hindi language, this process is performed by finding out the space separation and commas between the words. After that a list of words is formed and are called tokens. And this list is send for next level processing.

3. Stop-Words Elimination:

Generally, maximum commonly used words are referred to as stop words and they are not required for the text and hence are removed in this stage.

So in this way these common words should be eliminated from the original text because if we do not perform at this stage then the weight of these words grows maximum as can also affect the final output. From having prior observations that each Hindi textual content statistics has minimal 25% to 35% or maybe extra stop words. A Sample of stop-words are "के", "है", "और", "नहीं" etc.

b. Processing phase:

Processing phase is the most important phase in text summarization in any respective language. In this phase, feature value for each sentence is determined. And on bases of certain functions the final summary is being generated.

Stanza

Stanza is an open-source Python NLP toolkit which had a huge variety of sixty six human talking languages.). Stanza toolkit entails a language-agnostic completely neural pipeline for textual content examining, it has tokenization, multiword token expansion, lemmatization, part-of speech and morphological function tagging, dependency parsing, and named entity recognition (NER).

4. Design and Architecture:

Stanza have of two components:
1. Fully neural multilingual NLP pipeline,
2. Python client interface to the Java Stanford Core NLP software.

4.1 Neural Multilingual NLP Pipeline:

Stanza's neural pipeline has models that can perform operation from tokenizing raw text to accomplishing syntactic examination on complete sentences.

Segmentation and Tokenization:
Tool performs segmentation and then tokenization and makes a list respectively, Stanza combines tokenization and sentence segmentation from given input into one module and is additional named as tagging problem over character sequences, where the model guesses whether a given character is the end of a token, end of a sentence, or end of a multi-word token.

Multi-word Token Expansion:
It is obtained with associate degree ensemble of a frequency lexicon and a neural sequence-to-sequence model, to make sure that regularly ascertained expansions within the training set are invariably robustly enlarged whereas maintaining flexibility to model unseen words statistically.

POS and Morphological Feature Tagging:
Stanza assigns it part of speech to every word mentioned in the list, and observes its universal morphological features.

Lemmatization:
Stanza also performs lemmatization on every word in a sentence to retrieve its recognized form. Lemmatization is obtained as a collaboration of a dictionary-based lemmatizer and a neural seq2seq lemmatizer.

Dependency Parsing: Here every word in the sentence is allotted a syntactic head as Stanza parses every sentence of its syntactic structure.

Named Entity Recognition (NER):
Stanza also recognizes named entities in the given input in order to generate the required text summary.

4.2 Core NLP Client:
Stanford's Java Core NLP toolkit offers a collection of NLP tools but the drawback is that they are not smoothly accessible in python language, which is the majority choice of data scientists. But by applying certain methods we can make it flexible for python use too.

Spacy

Spacy is used for advanced natural language processing and is also open source so anyone can use it within just a few clicks. Spacy performs excellent for named entity recognition As time passes developers made spacy to train on more than one language especially in version 2.0 and after.

5. Implementation of Spacy

Its implementation phrase in python programming language consists of the following points.

a. Import libraries like (numpy, pandas, spacy, sklearn, string),
b. Providing input document,
c. Defining boundaries and performing segmentation,
d. Performing Tokenization,
e. Performing TF-IDF functionality,
f. Analysing top sentences,
g. Generating final summary.

6. Performing Comparison

In this phase we are going to perform a comparison of pre-processing time of both the tools, for this we considered 10 different Hindi Text Documents of size starting from 10,000 to 1,00,000 words. So, in this way we have 10 different observations and are quite sufficient to test whose pre-processing time is fast.

The processing of comparison involves, firstly import the respected dataset into the program, then for computing pre-processing time we imported time python library. We implemented this function at starting and end of the preprocessing step in order to compute the pre-processing time. After that separate readings of the text is taken for both the tools starting from ten thousand to one lakh word length. The pre-processing time is analysed in units of seconds i.e. the time required in the execution of pre-processing of the various text documents. After performing this operation on the entire dataset of text documents the following readings are found as shown in the table mentioned below:

Table 1. Comparison of pre-processing time in Stanza & Spacy

Text size (words)	Stanza (in sec)	Spacy (in sec)
10000	8.906458	0.049776
20000	19.81356	0.062253
30000	24.997198	0.081768
40000	32.280505	0.109722
50000	38.622927	0.127685
60000	46.119452	0.153165
70000	55.102151	0.171214
80000	63.901981	0.201384
90000	74.714428	0.221147
100000	86.042275	0.246014

Fig 5.1: Comparison Graph

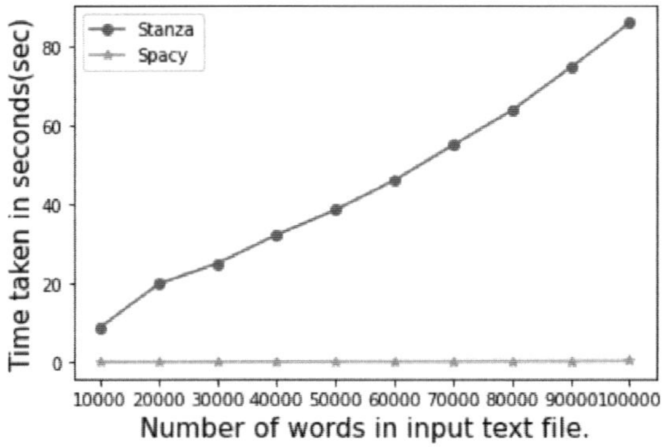

The results may slightly vary on systems with different configurations of hardware.

7.Conclusion

The final conclusion of this paper is that the pre-processing time of Stanza is very high as compared to Spacy, as it is clear from the above observation. But we can't forget that Stanza supports 66 languages whereas Spacy supports only 61 languages as of now. So, if the language you want to work on is supported by Spacy then it's better to go with spacy as it performs about 300 times faster as compared to Stanza. The pre-processing time in stanza increases with increase in the number of words in the input text. Both have their merits and demerits depending upon the place they are used but it's a better option to go with Spacy wherever it is applicable.

References

[1] Hans Peter Luhn. "The automatic creation of literature abstracts," IBM Journal of research and development, 2(2):159–165, 1958.

[2] P. B. Baxendale, "Machine-made Index for Technical Literature -An Experiment,"Journal of Research and IBM Development, vol. 2, no. 4, pp. 354-361, October 1958.

[3] Harold P Edmundson. "New methods in automatic extracting," Journal of the ACM (JACM), 16(2):264–285, 1969.

[4] Julian Kupiec, Jan Pedersen and Francine Chen, "A trainable document summarizer," Proceedings of the 18th annual international ACM SIGIR conference on Research and development in information retrieval, ACM, pp. 68-73, 1995.

[5] Gupta Vishal and Gurpreet Singh Lehal. "A survey of text summarization extractive techniques." Journal of Emerging Technologies in Web Intelligence, vol. 3, pp.258-268, 2010.

[6] Chin-Yew Lin, "Training a selection function for extraction," Proceedings of the eighth international conference on Information and knowledge management, ACM, pp. 55-62, 1999.

[7] M.J Conroy and O'leary, "Text summarization via hidden markov models," In Proceedings of SIGIR '01, pp.406-407, 2001.

[8] Md Haque, Suraiya Pervin and others, "Literature Review of Automatic Single Document Text Summarization Using NLP," International Journal of Innovation and Applied Studies, vol.3, no.3, pp. 857-865, 2013.

[9] Ladda Suanmali, Mohammed Salem Binwahlan and Naomie Salim, "Sentence features fusion for text summarization using fuzzy logic," Ninth International.

[10] Li Chengcheng, "Automatic Text Summarization Based On Rhetorical Structure Theory," International Conference on Computer Application and System Modeling (ICCASM), vol. 13, pp. 595-598, October 2010.
[11] S. Harabagiu and F. Lacatusu, "Using Topic Themes for MultiDocument Summarization," ACM Trans. Information Systems, vol. 28, pp. 13:1-13:47, July 2010.
[12] R. Varadarajan, V. Hristidis, and T. Li, "Beyond Single-Page Web Search Results," IEEE Trans. Knowledge and Data Eng., vol. 20,no. 3, pp. 411-424, Mar. 2008.
[13] E. Lloret, A. Balahur, M. Palomar, and A. Montoyo, "Towards Building a Competitive Opinion Summarization System: Challenges and Keys," Proc. Human Language Technologies: The 2009 Ann. Conference of the North Am. Ch. Assoc. for Computational Linguistics, Companion Vol.: Student Research Workshop and Doctoral Consortium, pp. 72,77, 2009.
[14] J.G. Conrad, J.L. Leidner, F. Schilder, and R. Kondadadi, "Query-Based Opinion Summarization for Legal Blog Entries," Proc. 12th Int'l Conf. Artificial Intelligence and Law, pp. 167- 176, 2009.
[15] H. Nishikawa, T. Hasegawa, Y. Matsuo, and G. Kikui, "Opinion Summarization with Integer Linear Programming Formulation for Sentence Extraction and Ordering," Proc. 23rd Int'l Conf. Computational Linguistics: Posters, pp. 910-918, 2010.

Indoor human occupancy detection using Machine Learning classification algorithms & their comparison

Disha Giri , Sakshi Shreya , Pushpanjali Kumari , Richa Yadav
Department Of Electronics and Communication Engineering
Indira Gandhi Delhi Technical University For Women, Delhi - 110006

Abstract- This paper presents a method to determine the occupancy in a room with the help of datasets whose data was collected from different sensors and using different ML algorithms. Features representing the occupancy level and the relative changes are taken from different sensors: Humidity sensors, light sensors,temperature sensors carbon dioxide (CO2) sensors. Different classification algorithms which are used for detection of occupancy are: Naive Bayes, classification via regression, random forest, simple logistic, multiclass classification, decision table. The result of the classifier can be classified in the class called - binary-class (which refers to the presence or absence of the person).

***Keywords*- Machine learning, weka, classification algorithms.**

1. Introduction

Business and private structures have gotten one of the significant energy customers in individuals' day by day life. Customarily, structures are molded with fixed timetables, for example beginning warming/cooling from early morning till late at night during non-weekend days and having greatest inhabitants during working hours. It is seen that such strategies may prompt critical waste in energy utilization, on the grounds that the warming/cooling and ventilation level set is independent of the genuine inhabitant's level [3]. Thus, it is very important to have methods which can help to detect the occupancy as it can be used for several purposes. In the present technological world, in automation systems detection of occupancy for different reasons can be a crucial job. It can be of immense use and most importantly its potential use in controlling electrical systems and devices like lighting, ventilation and air-conditioning which would be a great application.

There are several applications of Artificial Intelligence. One of which is Machine learning. It gives any work the strength to automatically find out and improve from previous experience and results without the need of explicit programming. It concentrates on the development of various other programs that can access data to train and use it to learn on their own through iterating processes and finally test themselves. Machine learning algorithm-based occupancy detection in this present time is very useful for different applications. In this paper we are using six machine learning classification algorithms which are Naive Bayes, classification via regression, random forest, simple logistics, multiclass classification and decision table. These classification methods are widely utilized and popular among different types of machine learning algorithms. The software used for implementing these algorithms is Weka. One of the several reasons for using weka is that one can experiment with their data set by applying various algorithms to know which model gives more accuracy quickly. In this paper these six classification algorithms are used to detect occupancy as well as air quality index. The main purpose of this paper is to compare the results obtained by different algorithms which comprises several factors including TP rate, FP rate etc in order to obtain the maximum accuracy. The results obtained were mainly focused on increasing the accuracy so that proper analysis could be done on them which can be further used for different purpose.

The progression of the paper is coordinated as follows: Section II incorporates the writing review of the current work. Section III includes the execution and wordings of the procedure and the related work. Section IV incorporates results and conversation of inhabitant's discovery and Section V incorporates the end and the future work.

2. Literature Review

Occupancy detection and its use has been studied and applied at various levels and have provided valuable information in the past. The related work done on the occupancy detection and smart building management systems using different methodologies and tools and frameworks have been listed down below in the given table.

Reference No.	Application	Hardware	Sensors	Techniques	Conclusion
[1]	Indoor Occupancy Detection	LORA	Temperature sensor;sound pressure sensing, light sensor; humidity, pressure, temperature sensor; a PaPIR motion sensor	IoT, Machine learning, neural network pattern recognition	The consequences of IoT sensor estimations with exactness up to 94.6% and 91.5% for the double (presence or absence of people) and multiclass issues, separately [1].
[2]	Indoor environment monitoring system	LoRa Semtech transceivers	Temperature; sound, pressure; a digital UV & ambient light sensor[2]; humidity, pressure.CO2 Volatile Organic Compounds (VOCs),	IoT	The results show that most of the sent messages are gotten by in any occasion 3 gateways,therefore having low group goof rates
[3]	Smart building; occupancy detection	-	CO2 sensors, PIR RH sensors, air speed sensors and globe thermometer [3].	Machine Learning/An Autoregressive Hidden Markov Model (ARHMM)	The result shows that the ARHMM method/technique has an average accuracy of 80.78%.

[4]	Occupancy estimation to set energy efficiency	-	PIR, CO2 sensors, RH sensors	Sensing; Processing; Statistical learning; Machine Learning	It provides energy efficiency up to 50%.
[5]	Building indoor occupancy, energy efficiency	-	Infrared (PIR) detection, and CO2 sensors. pressure sensors humidity, motion detection.	Direct discriminant examination (LDA), grouping and relapse trees & random forest (RF)models detecting as a substitute[13].	Controlled field investigation is guided in an indoor space to present that the proposed model can impersonate the CO2 assessments given the latent release rates [13].
[6]	Smartenviro n- ments,Indoor Occupancy Detection	RFID, Bluetooth	temperature, light intensity, noise, and CO2	Internet of Things, machine learning, data analysis	Examines techniques to anticipate the inhabitants of indoor conditions to settle on improved choices about ecological conditions
[7]	Analysis and estimation of occupancy in buildings	-	CO2, passive infrared (PIR), video, sound, badge counters,RFID sensors	Sensor-utility network (SUN) method	The outcomes from the extended ventilation examination affirm that SbP can spare 55% of all out ventilation contrasted and the traditional fixed timetable ventilation system [13].
[8]	Occupancy estimation	-	Ultrasound or PIR sensors, pressure or movement sensors [14].	Machine learning classifiers, Low pixel count thermal imaging technology	The paper checks the exhibition of an inhabitants put together assessment framework based with respect to a warm identifier cluster[14]

[9]	Occupancy Sensing	8 MHz microcontroller,Chipc on radio	PIR, and Grid-Eye Sensor	ML classifiers-artificial neural networks (ANN), K-nearest neighbors, (KNN), linear regression (LR)	This occupancy based monitoring system presents that 25.0% of energy savings are possible with occupant based conditioning methodologies.
[10]	Environmental Sensing And Location estimation	ATmega1284 microcontroller	A low power camera Cyclopes	Machine Learning - closest distance Markov Chain (CDMC), Blended Markov Chain	On examination of tests delivered by model utilizing different strategies shows that the model catches the inhabitant's elements precisely.
[11]	Occupancy system for reduc -tion of HVAC	ATmega128L microcontroller	A low power camera Cyclopes	ML -closest distance Markov Chain (CDMC), Blended Markov chain	The system effectively reduces the HVAC energy levels.
[12]	Occupancy Detection in Open offices	IP video camera network	iLON CO2 sensor	Data Mining-Single Object Access Protocol (SOAP) ML-Accelerated Statistical Learning (ASL)	The highlights from different sensors increment connections further, and over 90% data gain is obtained when at any rate 6 of the most educational highlights are utilized.

3. Implementation

For the implementation of the machine learning algorithms, we have used Weka software. Weka has a vast collection of visualization tools and algorithms for data analysis and predictive modelling, alongside graphical user interfaces for simple and straightforward access to those functions. It has tools for performing different purposes such as classification, data pre-processing, regression, clustering, association rules, and visualisation. We can experiment with our data set by applying various algorithms to know which model gives more accuracy quickly.

A **Naive Bayes** classifier assumes the presence of a specific feature during a class which is unrelated to the presence of the other feature. It is a ML model that is used for large volumes of data (millions of data). It is good for sentimental analysis.

Classification Via Regression: Principle task that empowers the utilization of arrangement calculations in relapse issues is the change of used on them.

Decision Table may be graphical thanks to map conditions against actions**.** It expresses the fact that certain actions are to be performed only if certain conditions are satisfied. Tabular form of conditions and actions can include more than one 'or' condition.
In this paper while discussing about different classification algorithms, we have found out accuracy of different algorithms on the basis of different factors:

A **confusion matrix** is a technique or method which is used in machine learning summarizing the performance of a classification algorithm. Classification accuracy in AI alone can be deceiving if there are an inconsistent number of perceptions in each class or if there are multiple classes in the dataset. Confusion matrices are one of the significant pieces of ML that are utilized to imagine significant prescient investigation like review, particularity, exactness, and accuracy. bunch of nonstop qualities into a bunch of spans which is what is utilized in this technique.

Random Forest is a regulated learning grouping calculation which might be utilized for both arrangements just as relapse. By and large, it is chiefly utilized for grouping issues [16].

Simple Logistic is an administered learning order calculation used to anticipate the likelihood of an objective variable. The idea of the objective variable is dichotomous, which implies there would be just two potential classes Confusion matrices are viewed as significant ML since they have given direct correlations of qualities like False Positives, True Positives, True Negatives and False Negatives.

The standard methodology is to utilize a solitary worth measurement to diminish every grid into one worth, and afterward to look at the measurement esteems. All in all, to analyse M1 and M2, we just look at f(M1) and f(M2), where work f is the single worth measurement that are either 0 or 1. [16]. This gives us an all-encompassing perspective.

In **Multi class classification**, we train a classifier on how well our arrangement model is using our training data and use a classifier for performing and what sorts of mistakes classifying the new dataset. In this splitting of it is making. It is used for assessing the dataset takes place into training and test the exhibition of an order model. datasets and then several other algorithms are:

In ML, **truth positive rate (TP)**, also mentioned sensitivity or recall, is employed to live the share of actual positives which are correctly identified.
In ML, A **false positive rate (FP)** is an outcome where the model incorrectly predicts the positive class, that is it is an accuracy metric that can be measured on a subset of machine learning models. In classification, pattern recognition, **Precision** is that, the fraction of relevant instances among the retrieved instances, while **Recall** is that, the fraction of the entire amount of relevant instances that were actually retrieved.

The **F-measure (Fbeta)** is calculated because the mean of precision and recall, giving each an equivalent weighting. It allows a model to be evaluated taking both the precision and recall into account using a single score.

IOP Conf. Series: Materials Science and Engineering 1110 (2021) 012020 doi:10.1088/1757-899X/1110/1/012020

The **Matthews correlation coefficient** (**MCC**) also called phi coefficient is used in ML as a measure of the quality of binary (two-class) classifications.

A ROC zone (collector working trademark territory) is a graphical portrayal which shows the presentation of a grouping model at all characterization limits. This bend plots two boundaries: True Positive Rate. Bogus Positive Rate.
Another region which can be utilized instead of a ROC bend is a Precision-Recall bend (PRC). It is utilized less as often as possible than ROC bends yet as we will see PRC may be a superior decision for imbalanced datasets.

Steps followed for obtaining the results: -

1) Open WEKA explorer.

2) Select the training dataset from the "Open File" under the Preprocess Tab option.

3) Go to the "Classify" tab for classifying the Unclassified data.

4) Click on the "Choose" button. From this, select the classifiers you want to apply.

5) In the test options select "Use training set" to train the model first .

6) Click on the "Start" button.

7) After that, select "Supplied test set" and provide the test data set.

8) Click on the "Start" button.

4. Results – Occupancy Analysis

SNo.	Algorithm	Accuracy	TP Rate	FP Rate	Precision	Recall	F-Measure	MCC	ROC Area	PRC Area
1.	Naive Bayes	98.3491%	0.983	0.013	0.984	0.983	0.984	0.952	0.995	0.989
2.	Classification Via Regression	92.8015%	0.928	0.170	0.927	0.928	0.927	0.778	0.846	0.896
3.	Decision Table	75.5127%	0.755	0.252	0.820	0.755	0.774	0.430	0.910	0.935
4.	Random Forest	93.8372%	0.938	0.120	0.939	0.938	0.938	0.815	0.990	0.990
5.	Simple Logistic	99.0874%	0.991	0.006	0.991	0.991	0.991	0.973	0.996	0.995
6.	Multi Class Classifier	97.4979%	0.975	0.074	0.975	0.975	0.975	0.924	0.945	0.994

a) Temperature value Vs Frequency of Occurrence of Occurrence that Temperature value

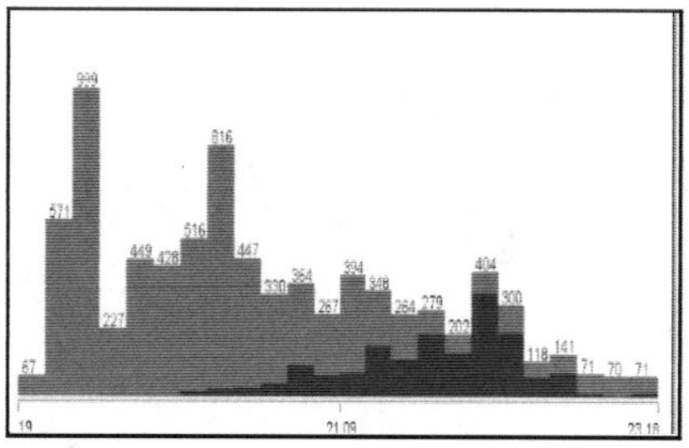

Temperature

b) Humidity value Vs Frequency of of that Humidity value

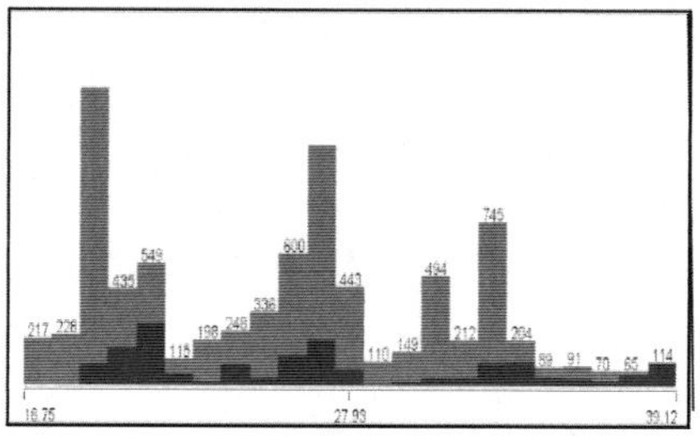

Humidity

c) Light intensity value Vs Frequency of occurrence of that Light intensity value

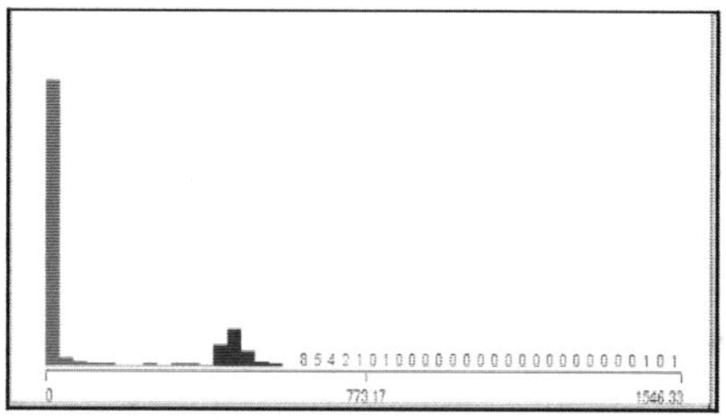

Light Intensity

d) CO2 value and Frequency of occurrence of that CO2 value

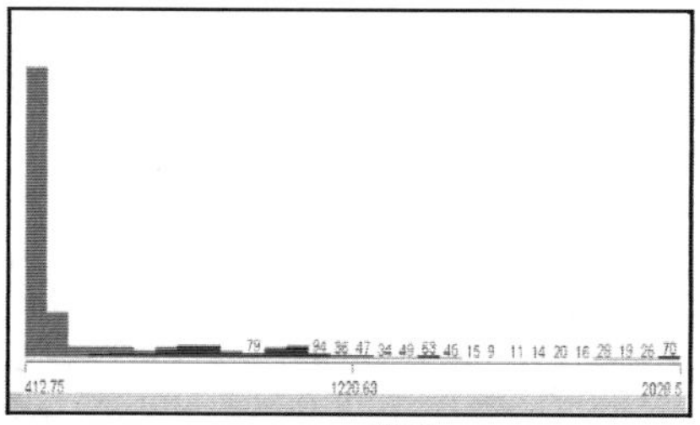

CO2

c) HumidityRatio value Vs Frequency of occurrence of that HumidityRatio value

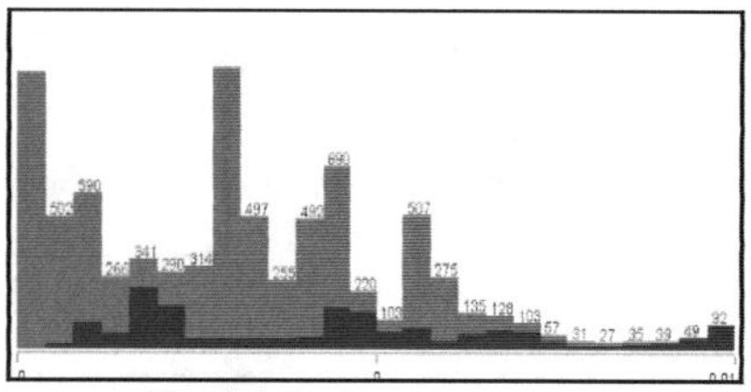

HumidityRatio

d) Occupancy

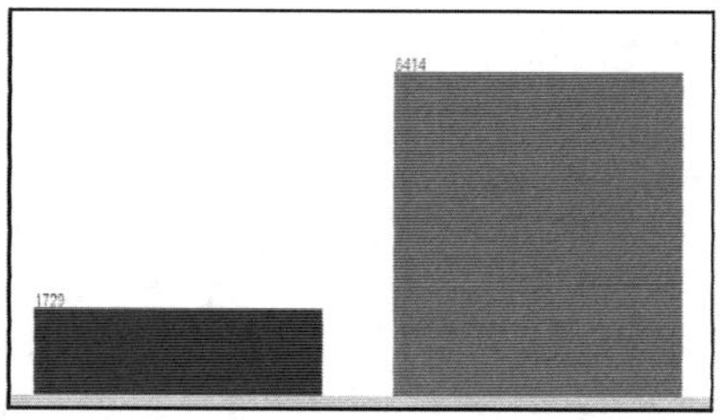

Occupancy

→ **Denotes absence of occupants**

→ **Denotes presence of occupants**

The above results in the form of images are the plots of different parameters: a) temperature b) humidity c) light d) CO2 e) humidity ratio f) Occupancy which are obtained from the test dataset plots on Weka. The accuracy and other important result parameters are tabulated above for occupancy detection analysis. The graphical representation of test datasets as simulated and pre-processed on Weka are also displayed. The red color symbolizes the absence of the occupant whereas the blue color indicates the presence of the occupant. The best accuracy is achieved for simple logistics (99.0874 %) and the worst by the decision table (75.5127 %)

5. Conclusion & Future Work

This paper speaks to a technique to identify inhabitants designs in a shut room. A strategy for inhabitants' discovery, upgraded by data hypothesis and factual learning is illustrated. The inhabitants' datasets are investigated utilizing calculations, for example, Naive Bayes, arrangement through relapse, irregular backwoods, basic strategic, multiclass order, choice tables. Results from the examination have indicated that Simple logistics has the best presentation in catching the space inhabitants' elements among the six calculations, with an exactness of 99.0874%. Such exactness shows its potential in improving structure energy productivity and adding to a greener climate.

Future work incorporates gathering the necessary information from building up the sensor hub organization, however more significantly the investigation and use of the gathered information to streamline indoor climate, both to spare water, warmth, and power, yet in addition improve the government assistance of representatives and visitors. From an exploration point of view, a broader start to finish execution investigation will be performed, while the organization is advanced and enhanced to gather more information or work with various designs, for example, actualizing our own AI grouping calculation to improve the inhabitant's discovery.

References

[1] Adeogun R, Rodriguez I, Razzaghipour M, Berardinelli G, Christensen P H, and Mogensen P E,*Indoor Occupancy Detection and Estimation using Machine Learning and Measurements from an IoT LoRa-based Monitoring System*, 2019.
[2] Rodriguez I, Lauridsen M, Vasluianu G, Poulsen A N, and Mogensen P, *The Gigantium Smart City Living Lab: A Multi-Arena LoRabased Testbed*, in 2018 15th International Symposium on Wireless Communication Systems (ISWCS), Aug 2018, pp. 1–6.
[3] Han Z, Gao R X, and Fan Z, *Occupancy, and indoor environment quality sensing for smart buildings*, in 2012 IEEE International Instrumentation and Measurement Technology Conference Proceedings, May 2012, pp. 882–887.
[4] Mulia M T, Supangkat S H, and Hariyanto N, *A review on building occupancy estimation methods*, in 2017 International Conference on ICT For Smart Society (ICISS), Sep. 2017, pp. 1–7.
[5] Jin M, Bekiaris N, Weekly K, Spanos C J, and Bayen A M, *Occupancy detection via environmental sensing*, IEEE Transactions on Automation Science and Engineering, vol. 15, no. 2, pp. 443–455, April 2018.
[6] Abade B, Abreu D, and Curado M, *A Non-Intrusive Approach for Indoor Occupancy Detection in Smart Environments,* sensors,vol. 18.
[7] Meyn S, Surana A, Lin Y and Oggianu S M, *A Sensor-Utility-Network Method for Estimation of Occupancy in Buildings*,in Conference on Decision and Control and 28th

Chinese Control Conference, Shanghai, 2009.
[8] Tyndall A, Cardell R and Keating A, *Occupancy Estimation Using a Low-Pixel Count Thermal Imager*, IEEE Sensors Journal, vol. 16, no. 10, pp. 3784-3791, 2016.
[9] Beltran A, Erickson V L, and Cerpa A E, *Thermosense: Occupancy thermal based sensing for hvac control*, in Proc. 5th ACM Workshop Embedded Syst. Energy-Efficient Buildings, 2013, pp. 1–8.
[10] Jin M, Zou H, Weekly K, Jia R, Bayen A M, and Spanos C J, *Environmental sensing by wearable device for indoor activity and location estimation*, in Proc. 40th Annu. Conf. IEEE Ind. Electron. Soc. (IECON), Oct. 2014, pp. 5369–5375
[11] Erickson V L, Carreira-Perpinan M A, Cerpa A E, *OBSERVE: occupancy-based system for efficient reduction of HV AC energy*, Proc. International Conference on Information Processing in Sensor Networks, pp. 258-269, Chicago, lllinois, USA, April. 2011
[12] Lam K P, Hoynck M, Zhang R, Andrews B, Chiou Y S, Dong B, Benitez D, *Information-theoretic environmental features selection for occupancy detection in open offices*, Proc. 11th International IBPSA Conference, pp.1460-1467, Glasgow, Scotland, July 2009
[13] https://bayen.eecs.berkeley.edu/sites/default/files/journals/Occupancy_Detection.pdf
[14] http://ash.id.au
[15] http://docplayer.n.et
[16] https://morioh.com/p/51b5c19afa26
[17] Agarwal Y, Balaji B, Dutta S, Gupta R, Weng T, *Duty cycling buildings aggressively: the next frontier in HVAC control*, Proc. International Conference on Information Processing in Sensor Networks. pp. 246-257. Chicago, Illinois, USA, April. 2011
[18] Wong L T, Mu K Wi, Shi K L, *Energy impact of indoor environmental policy for air-conditioned offices of Hong Kong*, Energy Policies, Vol. 36, pp. 714-721, 2008
[19] Pérez-Lombard L, Ortiz J, Pout C, A review on buildings energy consumption information, Energy and Buildings 40 (3) (2008) 394–398

AUTHOR INDEX

Institute of Physics
Dirac House, Temple Back
Bristol BS1 6BE UK

ISSN: 1757-8981
ISBN 978-1-7138-3400-7

CURRAN ASSOCIATES INC.
proceedings.com